Introduction: Genes, cognition and neuropsychiatry

Brita Elvevåg and Daniel R. Weinberger

Routledge
Taylor & Francis Group
LONDON AND NEW YORK

First published 2009 by Psychology Press

Published 2019 by Routledge
2 Park Square, Milton Park, Abingdon, Oxon OX14 4RN
52 Vanderbilt Avenue, New York, NY 10017

Routledge is an imprint of the Taylor & Francis Group, an informa business

ISBN 13: 978-1-84872-718-2 (hbk)

COGNITIVE NEUROPSYCHIATRY
2009, 14 (4/5), 261–276

Psychology Press
Taylor & Francis Group

Introduction: Genes, cognition and neuropsychiatry

Brita Elvevåg and Daniel R. Weinberger

Clinical Brain Disorders Branch, National Institute of Mental Health, National Institutes of Health, Bethesda, MD, USA

Numerous genes modulate dopamine and consequently prefrontal cortical function. Some of these genes – notably catechol-O-methyltransferase – have been shown to impact a variety of core cognitive processes that are dependent upon the prefrontal cortex. This demonstration that a single functional polymorphism can contribute so dramatically to individual differences heralds a new era for neuropsychiatry. Although enormous detail remains to be discovered about these various genes that regulate neurotransmitters, how they interact and how they affect brain systems, there is much excitement and promise regarding new neuropsychopharmacological possibilities. However, this new research program is magnitudes more complex than any enterprise embarked on hitherto and requires the development, validation and deployment of novel behavioural and neurophysiological phenotypes in order to unravel the pathologies within neural functional systems. This research foundation is essential if these genetic breakthroughs are to be translated to successful clinical agents.

Keywords: COMT; Dopamine; Prefrontal function; Phenotypes.

INTRODUCTION

Dopamine and neuropsychiatry: Evolving ideas

The dawn of the biological era of psychiatry was intertwined with a theory of the centrality of dopamine in the pathogenesis of a spectrum of psychiatric disorders and subsequent pharmacological interventions. Indeed, the dopamine hypothesis has been the most enduring theory concerning schizophrenia; it has been a central element in efforts to understand ADHD, and it has also been important in formulations of bipolar disorder

Correspondence should be addressed to Brita Elvevåg or Daniel R. Weinberger, Clinical Brain Disorders Branch, National Institute of Mental Health, National Institutes of Health, Building 10, Room 3C104, MSC 1379, Bethesda, MD 20892, USA. E-mail: elvevaagb@mail.nih.gov or weinberd@mail.nih.gov

This research was supported by the Intramural Research Program of the National Institute of Mental Health, National Institutes of Health.

http://www.psypress.com/cogneuropsychiatry DOI: 10.1080/13546800903126016

and depression. The relationship between the D_2 blocking actions of antipsychotic medications and their clinical benefit provided a logical foundation for the notion of dopamine overactivity – either presynaptically or postsynaptically – being the fundamental underlying mechanism for the presenting psychosis, and indeed recent research demonstrates more D_2 receptors as well as an increased level of dopamine in patients with schizophrenia (Abi-Dargham et al., 2000).

Despite the longevity of the dopamine hypothesis, it has justifiably been criticized for numerous reasons, notably seemingly poor correlation between antipsychotics' occupation of D_2 receptors and clinical efficacy, unwanted side-effects such as extrapyramidal signs (at $\approx$70% D_2 occupancy) and frequent incomplete resolution of the psychosis. Furthermore, there is much more to schizophrenia than the clinical presentation of psychosis. Specifically, cognitive deficits are widely recognized as being present prior to the florid illness onset, and remain despite moderately successful neuroleptic treatment, but the pathophysiology of this is unclear. Although it is indeed possible that aberrations in dopamine-mediated frontal-striatal loops could produce widespread cognitive dysfunction (Robbins, 1990), of current interest are evolving ideas that there are complex genetic mechanisms underlying dopamine modulation. Indeed, the idea is that a combination of susceptibility genes and other factors – including environmental – contribute to the complete picture of schizophrenia, namely the variables present at the level of clinical diagnosis as well as those evident at the level of "intermediate phenotypes". The phenotypes are intermediate in the sense that they bridge the gap between the effects of genes on cells and the emergent psychosis. Despite some dissent concerning the viability of this research methodology (Flint, 2007), a compelling case can be made for the need and promise in neuropsychiatry of intermediate phenotype research (Goldberg & Weinberger, 2004; Meyer-Lindenberg & Weinberger, 2006; Tan, Callicott, & Weinberger, 2008). Many of the intermediate phenotypes of interest, (e.g., working memory and executive function), have been shown to be inherited and "anomalies" are evident in unaffected siblings who undeniably are at increased genetic risk (Snitz, MacDonald, & Carter, 2006), and importantly are present prior to the florid onset of psychosis. Furthermore, these intermediate phenotypes are often more reducible and thus more amenable to systematic neurobiological research than phenomenology for example, and importantly in many cases the genetic associations show stronger penetrance at this intermediate phenotype level than at the level of clinical diagnosis (Tan et al., 2008). Very similar reasoning has been applied to reducing the complexity of genetic effects in other common complex medical disorders, such as adult onset diabetes and hypertension, where putative risk genes show much greater effects at intermediate phenotypes related to body mass

(Frayling et al., 2008) and sodium homeostasis (Newton-Cheh, Larson, & Vasan, 2009), respectively.

Dopamine is well documented as a major neurotransmitter in higher order behavioural and cognitive functions (Seamans et al., 2001; Seamans & Yang 2004) with the level of dopamine in the prefrontal cortex (PFC) being a critical modulator of many core cognitive processes (i.e., tuning these neuronal and circuit responses) (Levy & Goldman-Rakic, 2000; Robbins, 2000). Recent studies indicate that genetic factors affect dopamine flux in PFC. Thus, genes that regulate dopamine in PFC – specifically the catechol-O-methyltransferase (COMT) gene on chromosome 22q11 – are a crucial starting place in understanding the complex molecular genetic basis of cognitive function and dysfunction. COMT is an enzyme that plays a central role in the degradation of cortical dopamine (Tunbridge et al., 2004). A substitution of valine (Val) by methionine (Met) in the peptide sequence (commonly referred to as Val158Met) impacts the thermostability and reduces enzymatic activity more than half (Chen et al., 2004; Weinshilboum et al., 1999) such that the more stable Val allele will be associated with greater dopamine degradation and hence less synaptic dopamine than the less stable Met allele. This has surprisingly robust effects on prefrontal DA (Abi Darghum et al., 2008), subsequent prefrontal cortical physiology and prefrontally mediated cognition (Meyer-Lindenberg & Weinberger, 2006) (e.g., executive function and working memory – Aguilera et al., 2008; Bruder et al., 2005; Diaz-Asper et al., 2008; Egan et al., 2001; Mattay et al., 2003 – and attentional control – Blasi et al., 2005; Krabbendam et al., 2006; Winterer et al., 2006).

This direct association between a single functional polymorphism and behaviour although modest in absolute terms heralds a new era for explorations of the numerous levels of complexity between genes and cognition. Indeed, COMT has truly been a "soup to nuts"[1] story. It has provided insights into behavioural and neurophysiological phenotypes, contributed to the characterization of intra- and inter-regional neural systems underpinning cognitive behaviour, revealed the complexity of haplotype, gene–gene interaction, as well as the modulatory role of gene–environment interaction effects on cognition, demographic and developmental effects on the gene–cognition associations, and provided an early glimpse into how genes may interface between cognition and emotion (for a review, see Dickinson & Elvevåg, 2009). Last, but not least, COMT has highlighted both the urgency of translating this knowledge into practical pharmacogenetic agents as well as being a humbling reality-check of how much remains to be discovered.

---

[1] American idiom (1920s/1930s) conveying the meaning of "from beginning to end", derived from the description of a full course dinner.

Cognition and dopamine: Evolving probes (complex or simple?)

In moving beyond this relatively early stage of genetic dissection of cortical function, it is immediately apparent that the tools employed to measure cognition have many limitations. For example, the utility of examining the classic neuropsychological task – the Wisconsin Card Sorting Test (WCST) – further in terms of functional polymorphisms is probably of extremely limited value. Its use – and that of many other neuropsychological tests – although heuristic, has been opportunistic simply because there was a large amount of performance data available. However, these tests are complex and can be performed poorly and failed for many reasons, and indeed a generalized deficit profile for schizophrenia has often been reported, which may either be attributable to psychometric limitations of such tasks or due to the disease itself (Dickinson et al., 2008). Future cognitive genomic investigations clearly require more sophisticated psychometric assays. These are needed in order to parse precisely and segregate these underlying neural and cognitive processes, or at the very least restrict the options by which a person succeeds or fails on a task, and so uncover the effect of a functional polymorphism on functional and neurocognitive phenotypes.

The level of detail at the psychometric level will certainly need to be increased. Hitherto, most studies have focused on mean differences between groups of people, assayed variously as on average, slower or poorer performance. This is an oversimplification of behaviour patterns and unfortunate given that dopamine is considered to be the critical modulator of intra-individual variability (Servan-Schreiber, Printz, & Cohen, 1990). Indeed, it is intuitive that there is clinical value to observations of fluctuations in behaviour, such as variations in performance day-to-day or even trial-to-trial. For ease of experimental operationalization "average" performance has been the preferred metric, which is certainly a fairly good proxy when intra-individual variability in performance is minimal. However, this is not the case when there is considerable non-random intra-individual variability in which case subsequent conclusions are likely to be simplified or completely flawed. In this case, fluctuations in performance – whether they be within the same or across different tasks – reflect transient changes rather than more enduring changes as a function of learning. In the case of neurogenerative pathology and some brain disorders, there is already an impressive body of work that illustrates more intra-individual variability in cognitive functioning (for a review, see MacDonald, Nyberg, & Bäckman, 2006), as well as across the normal healthy lifespan (Williams et al., 2005).

Thus, a case could be made that very simple measures – analysed in great detail – would be beneficial in the current enterprise. In the case of simple decision making, cognitive science is still grappling with establishing the best model by which very simple decisions are made after a sufficient amount of noisy stimulus information has been accumulated (for a review, see Smith & Ratcliff, 2004). Understanding the time-course of simple decisions is an important building block to be established before building more complex structures. Examining the role of intra-individual variability in this process is likely to be critical, especially when individual differences are of interest as they are in "intermediate phenotype" research. For example, within-person trial-to-trial variability in perceptual speed has been shown to precede and predict greater cognitive decline in a large scale longitudinal study of elderly people (Lövdén et al., 2007). The value to the intermediate phenotype approach of examining individual variability is also nicely illustrated by more individual variability on an attention task in adolescents with two copies of a risk allele for ADHD (a 3' VNTR ten-repeat allele of the dopamine transporter gene) (Bellgrove et al., 2005). A recent illustration of the successful synthesis of theoretical computational neuroscience with an empirical genomic cognitive approach is an examination of intra-individual variabilities that underlie very basic perceptual 2-choice decision-making (Li et al., 2008). Information processing fluctuations were defined in terms of the standard deviation of each person's decision time distribution. In line with the idea that there is more efficient prefrontal dopamine modulation in met carriers (COMT val158met), the authors found less processing fluctuation (i.e., less intra-individual variation) in met homozygotes. To model the efficiency with which individuals accumulated evidence in order to make a decision, they fitted each person's decision time distribution to a popular "evidence accumulation model", namely the diffusion model which is a continuous stage random-walk sequential sampling model (for further explanation of such models, see Smith & Ratcliff, 2004). In line with predictions they found that met carriers were more efficient evidence accumulators, and specifically 10.4% of the variance in this (drift rate) was accounted for by COMT genotype. Importantly, they also evaluated the effect of the dopamine transporter gene (DAT1) – primarily expressed in striatum – which not surprisingly did not influence processing fluctuation of evidence accumulation rate. This experimental approach provides a novel framework to exploring the role that dopamine has in the temporal dynamics and quality of our integration of information in order to make very simple decisions.

Likewise at a neurophysiological level, intra-individual variability in percentage signal change across activation trials is likely to be associated with cognitive intra-individual variability. Intra-individual electrophysiological response variability in a frontal P300 component during an auditory oddball task has been shown to be positively associated with P300 amplitude, but inversely correlated with N-back working memory performance (Winterer et al., 2004). Further, exploring the role of COMT genotype in modulating this association, Val homozygotes displayed more prefrontal noise than Met carriers (but there was no difference between Met/Met and Val/Met groups). Such findings lend support to a critical role of dopamine in information processing, specifically by reducing the variability in signal and improving the signal-to-noise ratio. Such systematic examinations of intra-individual variability promise to create new behavioural and neurophysiological metrics that are useful proxies for neural processing.

It has been proposed that tonic stimulation of D1 receptors underlies the stabilization and maintenance of relevant information, and phasic stimulation of D2 receptors underlies the flexibility of processing information such as updating and manipulating new information (Durstewitz and Seamans, 2002; O'Reilly 2006; Seamans et al., 2001; Seamans & Yang, 2004; Williams & Goldman-Rakic, 1995). A mechanistic account of the effect of the COMTval158met polymorphism on these receptor specific effects suggests that whereas the Val allele is associated with increased phasic dopamine transmission and decreased tonic dopamine transmission, the Met allele is associated with an increase in tonic dopamine transmission and a decrease in phasic dopamine transmission. This would thus suggest that Val carriers will display poorer performance and efficiency in working memory and executive function tasks that tax the maintenance component of a task, and that Met carriers will be less efficient when rapid updating is required (Bilder et al., 2004). In line with this, Nolan et al. (2004) operationalized these demands in a response shifting task, and reported that the Met allele was beneficial for the operationalization of cognitive stability, but not cognitive flexibility.

Indeed, the "first-generation" psychometric tasks have been heuristic and provided important clues, but their original purpose ranged from tracking scholastic achievement to selecting soldiers for the military and generally parsed cognition in a rather simple sequential manner. The design of the subsequent "second-generation" cognitive tasks to parse genetics must be informed by a wide range of neurobiological sciences.

Genes, dopamine and individual differences: Even more complexity

Numerous factors have extremely complex effects on gene–cognition associations. Notably estrogen reduces COMT activity such that the effects of this genotype are generally different in women as compared to men, although the magnitude of this varies with age because of its rise and fall with the menopause (Harrison & Tunbridge, 2008). Indeed, some gene effects – notably COMT – are magnified in older people, with age being conceptualized as a shift leftwards – away from the functional optimum – towards the left-part of the inverted U-shaped dopamine signaling function (see Figure 1). Although likely an oversimplification of the complexity of aging, this conceptualization has been heuristic. For example, examining working memory and executive functioning in a large sample of young versus older adults, Nagel et al. (2008) found the Val disadvantage on the WCST and slower responding on a spatial working memory task but only among the older group. Interestingly, accuracy measures on this working memory test were not as sensitive to genetic effects as were latency measures. This is probably not surprising both for psychometric reasons inherent in their task as well as the sensitivity of average latency measures to subtle individual variations. Importantly the effect of COMT was modulated by a second polymorphism, namely Val66Met BDNF such that older people with two COMT val alleles and (at least) one BDNF met allele displayed longer latencies in performance. As the repertoire of candidate genes in researchers' toolkits increases it is likely that the interactions between genes, brain circuits and cognitive functions will be shown to be extremely complex, and require increasingly sophisticated computational and statistical data mining methods.

In a similar vein to findings of associations of COMT with differences in cognition, so too there are findings of the COMT polymorphism modulating affective dimensions, which may account for some individual variability in terms of vulnerability to stress and the subsequent development of anxiety

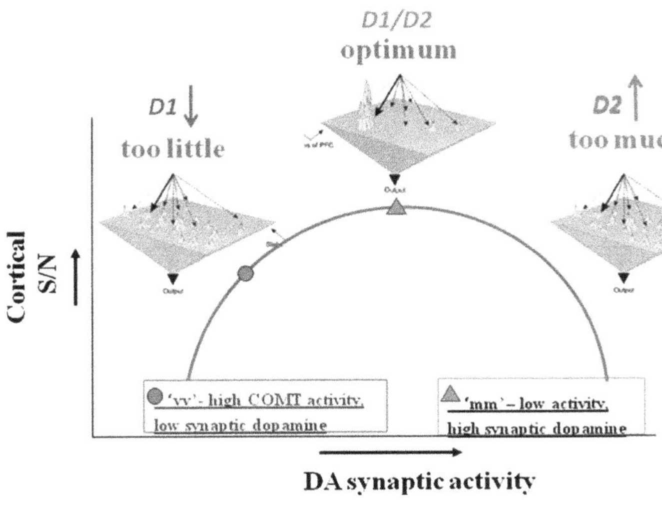

Figure 1. Dopamine signaling and prefrontal cortical function have an inverted U-shaped relationship. Axes illustrate cortical signal to noise as a function of dopamine synaptic activity. (Based on the work of Arnsten & Goldman-Rakic, 1986; Arnsten et al., 1994; Murphy et al., 1994, 1996a, 1996b, 1997; Williams & Goldman-Rakic, 1995; Verma & Moghaddam, 1996; Seamans et al., 2001; Vijayraghavan et al., 2007.) To view this figure in colour, please visit the online version of this issue.

disorders. As shown originally in monkeys, stress (noise) impairs delayed-response performance (i.e., moving one's hypothetical position on the inverted U-shaped response function), and interestingly can be ameliorated by blocking dopamine receptors (Arnsten & Goldman-Rakic, 1998). Not surprisingly, associations have been reported in humans between COMT and various temperamental traits such as extraversion, novelty seeking and reward dependence (Drabant et al., 2006; Reuter & Hennig, 2005; Tsai et al., 2004), and the negative and disorganization dimensions of schizotypy (Stefanis et al., 2004). Similarly, COMT has been considered a moderator of stress-induced psychosis (Stefanis et al., 2007). One possible neural mechanism for this is that the connectivity of brain systems may be different as a function of the COMT Val158Met genotype, with greater connectivity between the amygdala and orbital frontal cortex in met/met individuals that is associated with more rigidity and regimentation (as indexed by a correlation of a personality measure of novelty seeking behavior with *in vivo* examination of neural mechanisms involved during viewing of faces displaying negative emotions), a finding that lends itself to speculation of a increased vulnerability of "affective perseveration" (Drabant et al., 2006). In line with this, COMT met-allele carriers have been found to display an exaggerated startle reflex reaction to aversive stimuli (Montag et al., 2008), a result that fits well with the notion that the risk of developing anxiety disorders may be due to such genetic differences that underlie the neural regulation of fear.

Anxiety is not the only trait that has been linked to COMT. There are numerous reports of the high activity allele being associated with hyperactivity and impulsivity which have implications for attention-deficit/hyperactivity disorder (Halleland et al., 2009), but other studies have not found an association between COMT and impulsivity (Hallikainen et al., 2000). This story is likely to get progressively complex as impulsivity itself is a highly complex trait. Indeed, genetic variations in genes affecting dopamine signaling at a striatal level have been shown to be associated with impulsivity. Since impulsivity is related to reward sensitivity, ventral striatal reactivity is probably highly intertwined with any polymorphisms that modulate prefrontal cortex processing (see e.g., Forbes et al., 2009) due to cortico-striatal interaction (Schmack et al., 2008).

In terms of COMT and prefrontal cognition, a recent study on stress reactivity and cognition in mice argues for an evolutionary trade-off between cognition and emotion. Transgenic mice overexpressing a COMT-Val polymorphism (i.e., had increased COMT activity) displayed more errors on cognitive tasks but less stress responsivity and pain sensitivity (Papaleo et al., 2008). Interestingly, the inverse was true in that animals with genetically engineered reduced COMT activity had improved working memory performance but also increased stress reactivity, thus suggesting that whereas increased COMT activity may be protective in stress it also impairs cortically dependent function.

This idea of temperament rendering individuals more or less responsive to the environment is also illustrated by findings of higher scores of self-reported schizotypy in individuals with the high activity COMT (Avramopoulos et al., 2002). Further, a neuroimaging demonstration of this individual variability in regulation of affect-related processing examined attentional regulation in the face of emotional distraction. As might be expected, a Val loading of COMT val158met was associated with more frontal (left ventrolateral PFC and left orbitofrontal cortex) and bilateral parahippocampal activity (Bishop et al., 2006).

It is noteworthy however, that a major limitation of the studies discussed thus far is that only one functional polymorphism in COMT has been examined, but complex affective and cognitive functions are likely to be a result of interactions of multiple genetic loci (e.g., Tan et al., 2007). Nonetheless the questions examined and the experimental and statistical methods have been heuristic, at least theoretically. Indeed COMT has been a useful example of the ways that genes can be used to parse the interaction between cognition and individual differences such as affect. Naturally, the ultimate test of the utility of such scientific theories will be in the clinical – especially psychiatric – domain.

Genes, cognition and neuropsychiatry: Evolving clinical targets

Neuropsychiatry stands to benefit enormously from a better understanding of individual differences in terms of enduring cognitive deficits, but also in terms of how individuals characterised by genotype respond phenotypically to medication. The central role that COMT has in modulating dopamine and its direct impact on prefrontal cognition would seem to be fairly well established. Indeed, a proof of concept "challenge" study demonstrated that the dopaminetic amphetamine improved the poorer Val homozygotes prefrontal cortical function, as indexed behaviorally by a reduction in perseverative errors on the WCST and *in vivo* by increased "efficiency" on the N-back working memory task (Mattay et al., 2003). This improvement is attributable to genotypically reduced tone since it improved in Val homozygotes who presumably had less available synaptic dopamine. In contrast, amphetamine had deleterious effects on Met homozygotes' working memory efficiency. It is interesting to note that similar effects were also reported in the genetically altered COMT mice (Papaleo et al., 2008).

Subsequent pharmacogenetic studies have explored explicitly the effect of direct "interventions". Since cognitive dysfunction is a core and enduring feature of the schizophrenia syndrome and is not targeted or improved by current antipsychotics, clinical proof of concept intervention studies in this domain are extremely welcome. Although modest, these studies have found that patients homozygous for the COMT Met allele improved significantly their performance with amphetamine on the N-back working memory task, and those homozygous for the Val allele did not (Bertolino et al., 2004; Weickert et al., 2004; Woodward, Jayathilake, & Meltzer, 2007). These findings may be taken to suggest that Met homozygotes' working memory improved by means of a more "efficient" prefrontal cortex (i.e., less dorsolateral prefrontal activation) (for a review on imaging genomics as a method to assay response to antipsychotic treatment, see Blasi & Bertolino, 2006). An explicit way to examine the role of inheritance in inter-individual variation in drug responsivity is motivated by the idea that identifying phenotype differences can help improve drug efficacy and reduce the risk of therapeutic inefficiency or drug toxicity. Genetic variations in pharmacokinetics as well as pharmacodynamics can influence how a patient responds to a specific medication (for a review, see Weinshilboum & Wang, 2004, 2006), which offers promise of a more personalized treatment of carefully genetically calibrated dosing (for a review in psychiatric genetics, see Prathikanti & Weinberger, 2005).

One promising avenue is the development of cognitive-enhancing drugs, for treating cognitive deficits – specifically working memory – associated with schizophrenia (for reviews, see Apud & Weinberger, 2006, 2007). By specifically manipulating only cortical dopamine, tolcapone – a centrally active COMT inhibitor – represents a targeted pharmacological approach to improving cognitive function, by increasing dopamine mediated prefrontal signal to noise ratio (Apud & Weinberger, 2007). Indeed, especially in Val homoygotes, there are reports of marked improvements in working memory, executive function and prepulse inhibition behaviourally and *in vivo* in terms of PFC working memory "efficiency" (as assayed via fMRI) in healthy volunteers (Apud et al., 2007; Giakoumaki et al., 2008). Given the enduring nature and centrality of working memory deficits in schizophrenia COMT inhibition has been suggested as a candidate for ameliorating some of the cognitive deficits associated with the illness. At this point it is important to note that although these findings have been of great success scientifically, the humbling reality is that they have thus far been of limited clinical efficacy.

Future directions: Moving targets

It is clear that future progress will depend disproportionately on improving the characterization of phenotypes (Green et al., 2008), and much work remains to be done in terms of identifying clinically valuable and biologically relevant intermediate phenotypes. One possible avenue for this is the intra-individual variability approach discussed above. Another approach – discussed by Tan and colleagues (Tan, Callicott, & Weinberger, 2009) in this issue – is the charting of functional networks, systems and patterns of cortical functional connectivity (Meyer-Lindenberg & Weinberger, 2006; Meyer-Lindenberg et al., 2006). A more dynamical approach still, is the addition of real-time as a dimension so as to chart the fine-grained temporal nature of neural systems and their interaction, clearly necessary since dopamine is dynamic and thus eventually needs to be modeled dynamically. Indeed, with the increasing availability of technologies to measure the real-time activity of molecules and neural systems, so too is a need for exploitation of a more dynamical cognitive neuroscience approach (Rolls et al., 2008; Ward, 2002). The future success of cognitive neurogenetics – especially in terms of contributions to neuropsychiatry – will be dependent upon combining realistic time constants at all levels of cognitive neuroscience (molecules, neural systems and cognition). This is necessary for a truly dynamical systems approach that can systematically explore the roles that genes have in determining patterns of neural activity and connectivity that underlie higher-level cognitive processes. Modeling this dynamical process holds real promise of uncovering novel, clinically useful neurocognitive and neurophysiological phenotypes.

REFERENCES

Abi-Dargham, A., Rodenhiser, J., Printz, D., Zea-Ponce, Y., Gil, R., Kegeles, L. S., Weiss, R., Cooper, T. B., Mann, J. J., Van Heertum, R. L., Gorman, J. M., & Laruelle, M. (2000). Increased baseline occupancy of D2 receptors by dopamine in schizophrenia. *Proceedings of the National Academy of Science of the United States of America*, *97*, 8104–8109.

Aguilera, M., Barrantes-Vidal, N., Arias, B., Moya, J., Villa, H., Ibanez, M. I., Ruiperez, M. A., Ortet, G., & Fananas, L. (2008). Putative role of the COMT gene polymorphism (Val158Met) on verbal working memory functioning in a healthy population. *American Journal of Medical Genetics Part B Neuropsychiatric Genetics*, *147B*, 898–902.

Apud, J. A., & Weinberger, D. R. (2006). Pharmacogenetic tools for the development of target-oriented cognitive-enhancing drugs. *NeuroRx*, *3*, 106–116.

Apud, J. A., & Weinberger, D. R. (2007). Treatment of cognitive deficits associated with schizophrenia: Potential role of catechol-O-methyltransferase inhibitors. *CNS Drugs*, *21*, 535–557.

Apud, J. A., Mattay, V., Chen, J., Kolachana, B. S., Callicott, J. H., Rasetti, R., Alce, G., Iudicello, J. E., Akbar, N., Egan, M. F., Goldberg, T. E., & Weinberger, D. R. (2007). Tolcapone improves cognition and cortical information processing in normal human subjects. *Neuropsychopharmacology*, *32*, 1011–1020.

Arnsten, A. F. T., & Goldman-Rakic, P. S. (1986). Reversal of stress-induced delayed response deficits in rhesus monkeys by clonidine and naloxone. *Society of Neuroscience Abstracts*, *12*, 1464.

Arnsten, A. F., & Goldman-Rakic, P. S. (1998). Noise stress impairs prefrontal cortical cognitive function in monkeys: Evidence for a hyperdopaminergic mechanism. *Archives of General Psychiatry, 55*, 362–368

Arnsten, A. F., Cai, J. X., Murphy, B. L., & Goldman-Rakic, P. S. (1994). Dopamine D1 receptor mechanisms in the cognitive performance of young adult and aged monkeys. *Psychopharmacology (Berl), 116*, 143–151

Avramopoulos, D., Stefanis, N. C., Hantoumi, I., Smyrnis, N., Evdokimidis, I., & Stefanis, C. N. (2002). Higher scores of self reported schizotypy in healthy young males carrying the COMT high activity allele. *Molecular Psychiatry, 7*, 706–711

Bäckman, L., Nyberg, L., Lindenberger, U., Li, S. C., & Farde, L. (2006). The correlative triad among aging, dopamine, and cognition: Current status and future prospects. *Neuroscience and Biobehavioral Reviews, 30*, 791–807.

Bellgrove, M. A., Domschke, K., Hawi, Z., Kirley, A., Mullins, C., Robertson, I. H., & Gill, M. (2005). The methionine allele of the COMT polymorphism impairs prefrontal cognition in children and adolescents with ADHD. *Experimental Brain Research, 163*, 352–360.

Bertolino, A., Caforio, G., Blasi, G., De Candia, M., Latorre, V., Petruzzella, V., Altamura, M., Nappi, G., Papa, S., Callicott, J. H., Mattay, V. S., Bellomo, A., Scarabino, T., Weinberger, D. R., & Nardini, M. (2004). Interaction of COMT (Val(108/158)Met) genotype and olanzapine treatment on prefrontal cortical function in patients with schizophrenia. *American Journal of Psychiatry, 161*, 1798–1805

Bilder, R. M., Volavka, J., Lachman, H. M., & Grace, A. A. (2004). The catechol-O-methyltransferase polymorphism: Relations to the tonic-phasic dopamine hypothesis and neuropsychiatric phenotypes. *Neuropsychopharmacology, 29*, 1943–1961.

Bishop, S. J., Cohen, J. D., Fossella, J., Casey, B. J., & Farah, M. J. (2006). COMT genotype influences prefrontal response to emotional distraction. *Cognitive, Affective & Behavioral Neuroscience, 6*, 62–70

Blasi, G., & Bertolino, A. (2006). Imaging genomics and response to treatment with antipsychotics in schizophrenia. *NeuroRx, 3*, 117–130.

Blasi, G., Mattay, V. S., Bertolino, A., Elvevåg, B., Callicott, J. H., Das, S., Kolachana, B. S., Egan, M. F., Goldberg, T. E., & Weinberger, D. R. (2005). Effect of catechol-O-methyltransferase val158met genotype on attentional control. *Journal of Neuroscience, 25*, 5038–5045.

Bruder, G. E., Keilp, J. G., Xu, H., Shikhman, M., Schori, E., Gorman, J. M., & Gilliam, T. C. (2005). Catechol-O-methyltransferase (COMT) genotypes and working memory: Associations with differing cognitive operations. *Biological Psychiatry, 58*, 901–907.

Chen, J., Lipska, B. K., Halim, N., Ma, Q. D., Matsumoto, M., Melhem, S., Kolachana, B. S., Hyde, T. M., Herman, M. M., Apud, J., Egan, M. F., Kleinman, J. E., & Weinberger, D. R. (2004). Functional analysis of genetic variation in catechol-O-methyltransferase (COMT): Effects on mRNA, protein, and enzyme activity in postmortem human brain. *American Journal of Human Genetics, 75*, 807–821

Diaz-Asper, C. M., Goldberg, T. E., Kolachana, B. S., Straub, R. E., Egan, M. F., & Weinberger, D. R. (2008). Genetic variation in catechol-O-methyltransferase: Effects on working memory in schizophrenic patients, their siblings, and healthy controls. *Biological Psychiatry, 63*, 72–79.

Dickinson, D., & Elvevåg, B. (2009). Genes, cognition and brain through a COMT lens. *Journal of Neuroscience*, in press.

Dickinson, D., Ragland, J. D., Gold, J. M., & Gur, R. C. (2008). General and specific cognitive deficits in schizophrenia: Goliath defeats David? *Biological Psychiatry, 64*, 823–827

Drabant, E. M., Hariri, A. R., Meyer-Lindenberg, A., Munoz, K. E., Mattay, V. S., Kolachana, B. S., Egan, M. F., & Weinberger, D. R. (2006). Catechol O-methyltransferase val158met genotype and neural mechanisms related to affective arousal and regulation. *Archives of General Psychiatry, 63*, 1396–1406

Durstewitz, D., & Seamans, J. K. (2002). The computational role of dopamine D1 receptors in working memory. *Neural Networks, 15*, 561–572.

Egan, M. F., Goldberg, T. E., Kolachana, B. S., Callicott, J. H., Mazzanti, C. M., Straub, R. E., Goldman, D., & Weinberger, D. R. (2001). Effect of COMT Val108/158 Met genotype on frontal lobe function and risk for schizophrenia. *Proceedings of the National Academy of Science of the United States of America, 98*, 6917–6922.

Flint, J., & Munafo, M. R. (2007). The endophenotype concept in psychiatric genetics. *Psychological Medicine, 37*, 163–180.

Forbes, E. E., Brown, S. M., Kimak, M., Ferrell, R. E., Manuck, S. B., & Hariri, A. R. (2009). Genetic variation in components of dopamine neurotransmission impacts ventral striatal reactivity associated with impulsivity. *Molecular Psychiatry, 14*, 60–70.

Frayling, T. M., Timpson, N. J., Weedon, M. N., Zeggini, E., Freathy, R. M., Lindgren, C. M., Perry, J. R., Elliott, K. S., Lango, H., Rayner, N. W., Shields, B., Harries, L. W., Barrett, J. C., Ellard, S., Groves, C. J., Knight, B., Patch, A. M., Ness, A. R., Ebrahim, S., Lawlor, D. A., Ring, S. M., Ben-Shlomo, Y., Jarvelin, M. R., Sovio, U., Bennett, A. J., Melzer, D., Ferrucci, L., Loos, R. J., Barroso, I., Wareham, N. J., Karpe, F., Owen, K. R., Cardon, L. R., Walker, M., Hitman, G. A., Palmer, C. N., Doney, A. S., Morris, A. D., Smith, G. D., Hattersley, A. T., & McCarthy, M. I. (2007). A common variant in the FTO gene is associated with body mass index and predisposes to childhood and adult obesity. *Science, 316*, 889–894.

Giakoumaki, S. G., Roussos, P., & Bitsios, P. (2008). Improvement of prepulse inhibition and executive function by the COMT inhibitor tolcapone depends on COMT Val158Met polymorphism. *Neuropsychopharmacology, 33*, 3058–3068.

Goldberg, T. E., & Weinberger, D. R. (2004). Genes and the parsing of cognitive processes. *Trends in Cognitive Science, 8*, 325–335

Green, A. E., Munafò, M. R., Deyoung, C. G., Fossella, J. A., Fan, J., & Gray, J. R. (2008). Using genetic data in cognitive neuroscience: From growing pains to genuine insights. *Nature Reviews Neuroscience, 9*, 710–720

Halleland, H., Lundervold, A. J., Halmøy, A., Haavik, J., & Johansson, S. (2009). Association between Catechol O-methyltransferase (COMT) haplotypes and severity of hyperactivity symptoms in adults. *American Journal Medical Genetics part B Neuropsychiatric Genetics, 150B*, 403–410

Hallikainen, T., Lachman, H., Saito, T., Volavka, J., Kauhanen, J., Salonen, J. T., Ryynänen, O. P., Koulu, M., Karvonen, M. K., Pohjalainen, T., Syvälahti, E., Hietala, J., & Tiihonen, J. (2000). Lack of association between the functional variant of the catechol-O-methyltransferase (COMT) gene and early-onset alcoholism associated with severe antisocial behavior. *American Jorunal of Medical Genetics, 96*, 348–352

Harrison, P. J., & Tunbridge, E. M. (2008). Catechol-O-methyltransferase (COMT): A gene contributing to sex differences in brain function, and to sexual dimorphism in the predisposition to psychiatric disorders. *Neuropsychopharmacology, 33*, 3037–3045

Krabbendam, L., Isusi, P., Galdos, P., Echevarria, E., Bilbao, J. R., Martin-Pagola, A., Papiol, S., Castano, L., & Van Os, J. (2006). Associations between COMTVal158Met polymorphism and cognition: Direct or indirect effects? *European Psychiatry, 21*, 338–342.

Levy, R., & Goldman-Rakic, P. S. (2000). Segregation of working memory functions within the dorsolateral prefrontal cortex. *Experimental Brain Research, 133*, 23–32

Li, S.-C., Lindenberger, U., Chicherio, C., von Oertzen, T., Nagel, I., Sander, T., Villringer, A., Heekeren, H. R., & Bäckman, L. (2008). *Genetic predisposition for dopamine modulation of processing fluctuation and evidence accumulation.* Presentation at the Days of Molecular Medicine meeting, Stockholm, April 17–19, 2008.

Lövdén, M., Li, S. C., Shing, Y. L., & Lindenberger, U. (2007). Within-person trial-to-trial variability precedes and predicts cognitive decline in old and very old age: Longitudinal data from the Berlin Aging Study. *Neuropsychologia, 45*, 2827–2838.

MacDonald, S. W., Nyberg, L., & Bäckman, L. (2006). Intra-individual variability in behavior: Links to brain structure, neurotransmission and neuronal activity. *Trends in Neuroscience, 29*, 474–480.

Mattay, V. S., Goldberg, T. E., Fera, F., Hariri, A. R., Tessitore, A., Egan, M. F., Kolachana, B., Callicott, J. H., & Weinberger, D. R. (2003). Catechol O-methyltransferase val158-met genotype and individual variation in the brain response to amphetamine. *Proceedings of the National Academy Science of the United States of America*, *100*, 6186–6191.

Meyer-Lindenberg, A., & Weinberger, D. R. (2006). Intermediate phenotypes and genetic mechanisms of psychiatric disorders. *Nature Reviews Neuroscience*, *7*, 818–827.

Meyer-Lindenberg, A., Nichols, T., Callicott, J. H., Ding, J., Kolachana, B., Buckholtz, J., Mattay, V. S., Egan, M., & Weinberger, D. R. (2006). Impact of complex genetic variation in COMT on human brain function. *Molecular Psychiatry*, *11*, 867–877, 797.

Montag, C., Buckholtz, J. W., Hartmann, P., Merz, M., Burk, C., Hennig, J., & Reuter, M. (2008). COMT genetic variation affects fear processing: Psychophysiological evidence. *Behavioral Neuroscience*, *122*, 901–909.

Murphy, B. L., Roth, R. H., & Arnsten, A. F. T. (1994). The effects of FG7142 on prefrontal cortical dopamine and spatial working memory in rat and monkey. *Society of Neuroscience Abstracts*, *20*, 1018.

Murphy, B. L., Arnsten, A. F., Goldman-Rakic, P. S., & Roth, R. H. (1996). Increased dopamine turnover in the prefrontal cortex impairs spatial working memory performance in rats and monkeys. *Proceedings of the National Academy of Science of the United States of America*, *93*, 1325–1329.

Murphy, B. L., Arnsten, A. F., Jentsch, J. D., & Roth, R. H. (1996). Dopamine and spatial working memory in rats and monkeys: Pharmacological reversal of stress-induced impairment. *Journal Neuroscience*, *16*, 7768–7775.

Murphy, B. L., Roth, R. H., & Arnsten, A. F. (1997). Clozapine reverses the spatial working memory deficits induced by FG7142 in monkeys. *Neuropsychopharmacology*, *16*, 433–437.

Nagel, I. E., Chicherio, C., Li, S., von Oertzen, T., Sander, T., Villringer, A., Heekeren, H. R., Bäckman, L., & Lindenberger, U. (2008). Human aging magnifies genetic effects on executive functioning and working memory. *Frontiers Human Neuroscience*, *2*, 1.

Newton-Cheh, C., Larson, M. G., Vasan, R. S., Levy, D., Bloch, K. D., Surti, A., Guiducci, C., Kathiresan, S., Benjamin, E. J., Struck, J., Morgenthaler, N. G., Bergmann, A., Blankenberg, S., Kee, F., Nilsson, P., Yin, X., Peltonen, L., Vartiainen, E., Salomaa, V., Hirschhorn, J. N., Melander, O., & Wang, T. J. (2009). Association of common variants in NPPA and NPPB with circulating natriuretic peptides and blood pressure. *Nature Genetics*, *41*, 348–353.

Nolan, K. A., Bilder, R. M., Lachman, H. M., & Volavka, J. (2004). Catechol O-methyltransferase Val158Met polymorphism in schizophrenia: Differential effects of Val and Met alleles on cognitive stability and flexibility. *American Journal Psychiatry*, *161*, 359–361.

O'Reilly, R. C. (2006). Biologically based computational models of high-level cognition. *Science*, *314*, 91–94.

Papaleo, F., Crawley, J. N., Song, J., Lipska, B. K., Pickel, J., Weinberger, D. R., & Chen, J. (2008). Genetic dissection of the role of catechol-O-methyltransferase in cognition and stress reactivity in mice. *Journal of Neuroscience*, *28*, 8709–8723.

Prathikanti, S., & Weinberger, D. R. (2005). Psychiatric genetics – the new era: Genetic research and some clinical implications. *British Medical Bulletin*, *73–74*, 107–122.

Reuter, M., & Hennig, J. (2005). Association of the functional catechol-O-methyltransferase VAL158MET polymorphism with the personality trait of extraversion. *NeuroReport*, *16*, 1135–1138

Robbins, T. W. (1990). The case of frontostriatal dysfunction in schizophrenia. *Schizophrenia Bulletin*, *16*, 391–402.

Robbins, T. W. (2000). Chemical neuromodulation of frontal-executive functions in humans and other animals. *Experimental Brain Research*, *133*, 130–138.

Rolls, E. T., Loh, M., Deco, G., & Winterer, G. (2008). Computational models of schizophrenia and dopamine modulation in the prefrontal cortex. *Nature Reviews Neuroscience*, *9*, 696–709.

Schmack, K., Schlagenhauf, F., Sterzer, P., Wrase, J., Beck, A., Dembler, T., Kalus, P., Puls, I., Sander, T., Heinz, A., & Gallinat, J. (2008). Catechol-O-methyltransferase val158met genotype influences neural processing of reward anticipation. *Neuroimage*, *42*, 1631–1638.

Seamans, J. K., & Yang, C. R. (2004). The principal features and mechanisms of dopamine modulation in the prefrontal cortex. *Progress Neurobiology*, *74*, 1–58.

Seamans, J. K., Gorelova, N., Durstewitz, D., & Yang, C. R. (2001). Bidirectional dopamine modulation of GABAergic inhibition in prefrontal cortical pyramidal neurons. *Journal Neuroscience*, *21*, 3628–3638.

Servan-Schreiber, D., Printz, H., & Cohen, J. D. (1990). A network model of catecholamine effects: Gain, signal-to-noise ratio, and behavior. *Science*, *249*, 892–895.

Slifstein, M., Kolachana, B., Simpson, E. H., Tabares, P., Cheng, B., Duvall, M., Frankle, W. G., Weinberger, D. R., Laruelle, M., & Abi-Dargham, A. (2008). COMT genotype predicts cortical-limbic D1 receptor availability measured with [11C]NNC112 and PET. *Molecular Psychiatry*, *13*, 821–827

Smith, P. L., & Ratcliff, R. (2004). Psychology and neurobiology of simple decisions. *Trends in Neuroscience*, *27*, 161–168.

Snitz, B. E., MacDonald, A. W., & Carter, C. S. (2006). Cognitive deficits in unaffected first-degree relatives of schizophrenia patients: A meta-analytic review of putative endophenotypes. *Schizophrenia Bulletin*, *32*, 179–194.

Stefanis, N. C., Henquet, C., Avramopoulos, D., Smyrnis, N., Evdokimidis, I., Myin-Germeys, I., Stefanis, C. N., & Van Os, J. (2007). COMT Val158Met moderation of stress-induced psychosis. *Psychological Medicine*, *37*, 1651–1656.

Tan, H. Y., Chen, Q., Sust, S., Buckholtz, J. W., Meyers, J. D., Egan, M. F., Mattay, V. S., Meyer-Lindenberg, A., Weinberger, D. R., & Callicott, J. H. (2007). Epistasis between catechol-O-methyltransferase and type II metabotropic glutamate receptor 3 genes on working memory brain function. *Proceedings of the National Academy of Science of the United States of America*, *104*, 12536–12541.

Tan, H. Y., Callicott, J. H., & Weinberger, D. R. (2008). Intermediate phenotypes in schizophrenia genetics redux: Is it a no brainer? *Molecular Psychiatry*, *13*, 233–238.

Tan, H. Y., Callicott, J. H., & Weinberger, D. R. (2009). Prefrontal cognitive systems in schizophrenia: Towards human genetic brain mechanisms. *Cognitive Neuropsychiatry*, *14*(4/5), 277–298.

Tsai, S. J., Hong, C. J., Yu, Y. W., & Chen, T. J. (2004). Association study of catechol-O-methyltransferase gene and dopamine D4 receptor gene polymorphisms and personality traits in healthy young chinese females. *Neuropsychobiology*, *50*, 153–156

Tunbridge, E. M., Bannerman, D. M., Sharp, T., & Harrison, P. J. (2004). Catechol-O-methyltransferase inhibition improves set-shifting performance and elevates stimulated dopamine release in the rat prefrontal cortex. *Journal of Neuroscience*, *24*, 5331–5335

Verma, A., & Moghaddam, B. (1996). NMDA receptor antagonists impair prefrontal cortex function as assessed via spatial delayed alternation performance in rats: Modulation by dopamine. *Journal of Neuroscience*, *16*, 373–379

Vijayraghavan, S., Wang, M., Birnbaum, S. G., Williams, G. V., & Arnsten, A. F. (2007). Inverted-U dopamine D1 receptor actions on prefrontal neurons engaged in working memory. *Nature Neuroscience*, *10*, 376–384.

Ward, L. M. (2002). *Dynamical cognitive science*. Cambridge, MA: MIT Press.

Weickert, T. W., Goldberg, T. E., Mishara, A., Apud, J. A., Kolachana, B. S., Egan, M. F., & Weinberger, D. R. (2004). Catechol-O-methyltransferase val108/158met genotype predicts working memory response to antipsychotic medications. *Biological Psychiatry*, *56*, 677–682

Weinshilboum, R., & Wang, L. (2004). Pharmacogenomics: Bench to bedside. *Nature Reviews Drug Discovery*, *3*, 739–748.

Weinshilboum, R. M., & Wang, L. (2006). Pharmacogenetics and pharmacogenomics: Development, science, and translation. *Annual Review Genomics Human Genetics*, *7*, 223–245.

Weinshilboum, R. M., Otterness, D. M., & Szumlanski, C. L. (1999). Methylation pharmacogenetics: Catechol O-methyltransferase, thiopurine methyltransferase, and histamine N-methyltransferase. *Annual Review Pharmacology Toxicology*, *39*, 19–52.

Williams, G. V., & Goldman-Rakic, P. S. (1995). Modulation of memory fields by dopamine D1 receptors in prefrontal cortex. *Nature*, *376*, 572–575.

Williams, B. R., Hultsch, D. F., Strauss, E. H., Hunter, M. A., & Tannock, R. (2005). Inconsistency in reaction time across the life span. *Neuropsychology*, *19*, 88–96.

Winterer, G., Coppola, R., Goldberg, T. E., Egan, M. F., Jones, D. W., Sanchez, C. E., & Weinberger, D. R. (2004). Prefrontal broadband noise, working memory, and genetic risk for schizophrenia. *American Journal of Psychiatry*, *161*, 490–500.

Winterer, G., Musso, F., Vucurevic, G., Stoeter, P., Konrad, A., Seker, B., Gallinat, J., Dahmen, N., & Weinberger, D. R. (2006). COMT genotype predicts BOLD signal and noise characteristics in prefrontal circuits. *Neuroimage*, *32*, 1722–1732.

Woodward, N. D., Jayathilake, K., & Meltzer, H. Y. (2007). COMT val108/158met genotype, cognitive function, and cognitive improvement with clozapine in schizophrenia. *Schizophrenia Research*, *90*, 86–96.

COGNITIVE NEUROPSYCHIATRY
2009, 14 (4/5), 277–298

Psychology Press
Taylor & Francis Group

Prefrontal cognitive systems in schizophrenia: Towards human genetic brain mechanisms

Hao-Yang Tan, Joseph H. Callicott, and Daniel R. Weinberger

Clinical Brain Disorders Branch, Genes, Cognition and Psychosis Program, National Institute of Mental Health, National Institute of Health, Bethesda, MD, USA

Schizophrenia has complex genetic heritability. It is also genetically heterogeneous. To the extent that genes are associated with symptom constellations in schizophrenia, they do so by affecting the development and function of neural systems that mediate the expression of such diverse behavioral, cognitive and perceptual phenomena. The genetic mechanisms of human brain dysfunction remain to be well understood. "Imaging genetics" is an emerging field that attempts to integrate the basic biology of putative disease mechanisms with physiological correlates from the live human brain. Here, we review recent imaging genetics work on prefrontal brain systems associated with working memory and executive function – heritable traits relevant to schizophrenia. Starting with genetic variation in dopaminergic systems (e.g., COMT), we examined the modulation of prefrontal brain networks during active cognitive processing; there is also evidence that variation in the expression of dopamine-related downstream intra-cellular signaling molecules (e.g., AKT1) are implicated. Moreover, these genetic variants evidence epistasis on neuroimaging measures, lending further support to the conceptualization that non-additive combinations of multiple genes modulate active human cognitive brain mechanisms. The imaging genetics platform therefore could extend understanding of genetic mechanisms of human cognitive brain processes relevant to neuropsychiatric disease.

Keywords: Working memory; MRI; Dopamine; Genetics.

INTRODUCTION

Cognitive deficits in neuropsychiatric disorders like schizophrenia remain unsolved and account for significant morbidity (Green, 1996). Recent real-world clinical

Correspondence should be addressed to Hao-Yang Tan or Daniel R. Weinberger, Clinical Brain Disorders Branch, Genes, Cognition and Psychosis Program, National Institute of Mental Health Intramural Research Program, National Institutes of Health, 10 Center Drive, Room 4S235, Bethesda, MD 20892, USA. E-mail: tanh@mail.nih.gov or weinberd@mail.nih.gov

http://www.psypress.com/cogneuropsychiatry DOI: 10.1080/13546800903091665

trials have highlighted the need to improve the efficacy of antipsychotic medications in symptom management and patient acceptability (Jones et al., 2006; Lieberman et al., 2005), let alone their limited effects on improving cognitive deficits central to the illness (Goldberg & Weinberger, 1996). The clinical data, nevertheless, support a canonical finding that the brain dopamine system, the common target of all antipsychotic drugs, remains an important aspect of the treatment and likely pathophysiology of psychosis. However, these brain mechanisms need to be better understood, for example in examining how genes and their products impact active human brain function and disease-related cognitive impairments.

The genetic diathesis of schizophrenia has been well documented by classical twin, adoption and family studies (Gottesman & Shields, 1972; Kety, Rosenthal, Wender, & Schulsinger, 1976). What has remained challenging is the elucidation of specific genes associated with this complex and genetically heterogeneous neuropsychiatric condition. Promising results have emerged from recent high throughput whole genome association studies involving multiple centers and thousands of patients, including combining schizophrenia and bipolar disorders patients to examine common disease genetics (O'Donovan et al., 2008). Replicable signals may also emerge, accelerated with the emergence of multiple large datasets. These whole-genome association studies have not (yet) implicated genes directly related to dopamine function, but could identify novel targets and paradigm shifts. However, recent meta-analyses of genetic association studies have pointed, for instance, to a number of dopaminergic genes (e.g., DRD1, DRD2), among others, that could play significant roles in psychosis (Allen et al., 2008). Cognizant that the literature can partly be driven by bias (Ioannidis, 2005), we nevertheless contend that since the weight of evidence-based treatments for psychosis point largely to the dopamine system, upon which much basic science data has also accumulated, genes related to the dopamine system serves as a logical starting point for a review on the genetics and potential brain mechanisms of psychosis. A goal of this review is thus to examine a selection of recent developments that integrate human neuroimaging, cognition and genetics in the elucidation of putative genetic mechanisms of human prefrontal cognitive functions of relevance to schizophrenia.

"INTERMEDIATE PHENOTYPE" APPROACH IN IMAGING GENETICS

Genes encode proteins, and not psychopathology, be they hallucinations, delusions, sadness or anxiety. To the extent genes are implicated, for example with the constellation of symptoms we call schizophrenia or bipolar disorder, they do so by affecting the development and function of brain cells and neural systems that mediate the expression of such diverse behavioral and perceptual phenomena. Patients with neuropsychiatric disease have changes in cognition, brain function and structure. These brain abnormalities are also found more frequently in their unaffected siblings, including unaffected monozygotic twins, than in control subjects without such a family history. These various deviations therefore represent biologic expressions of increased genetic risk rather than a disease entity per se. Since these biologic changes are found in individuals carrying greater genetic risk but who do not manifest a DSM-IV diagnosis (Blackwood, St Clair, Muir, & Duffy, 1991; Callicott et al., 2003a; Cannon et al., 2000; Goldberg et al., 1995; Goldman et al., 2008; Toulopoulou et al., 2007), these brain changes are susceptibility-related phenotypes, intermediate between the cellular effects of susceptibility genes and manifest psychopathology (Meyer-Lindenberg & Weinberger, 2006; Tan, Callicott, & Weinberger, 2008a; Weinberger et al., 2001).

Such genetic links, particularly with quantifiable functional brain abnormalities that are also heritable, should facilitate the elucidation of their underlying neural mechanisms. This is a principal value of studying specific intermediate brain phenotypes and their genetic associations. These studies characterize, in the live human brain, the neural system mechanisms of the clinical genetic associations (Meyer-Lindenberg & Weinberger, 2006; Tan et al., 2008a; Weinberger et al., 2001). An illustrative example that we will develop in this review is prefrontal cortically mediated working memory, implicated in the cognitive abnormalities and dopaminergic brain processes of psychosis, and its putative genetic mechanisms.

PREFRONTAL CORTEX, WORKING MEMORY AND SCHIZOPHRENIA

Working memory is a limited capacity system that enables us to temporarily hold, update and work with relevant information; it underlies almost all higher-order thinking, language and goal-directed behavior (Baddeley, 2003). Critically engaging prefrontal cortical brain systems, working memory has been shown to be an important component underlying many cognitive deficits observed in schizophrenia (Goldberg, Weinberger, Berman, Pliskin, & Podd, 1987; Goldman-Rakic 1994; Silver, Feldman, Bilker, & Gur, 2003). An extensive body of functional imaging experiments is also consistent with prefrontal cortical physiological dysfunction in schizophrenia (e.g., see Reviews by Callicott & Weinberger, 1999; Manoach, 2003; Tan, Callicott, & Weinberger, 2007a). The recent literature has also clarified that the directionality of the imaging findings (i.e., too much or too little prefrontal activity) depends on the specific task paradigm utilized, disease-related differences in behavioral performance, capacity constraints of task load

(Callicott et al., 1999, 2000, 2003b; Manoach, 2003; Ramsey et al., 2002), and the engagement of a system of dysfunctional and compensatory circuits (Tan, Choo, Fones, & Chee, 2005; Tan et al., 2006). In particular, absent task performance as a confounder, the increased prefrontal activation and physiological inefficiency phenotype has often been observed with schizophrenia, and with increased genetic risk for schizophrenia (Callicott et al., 2003a; Glahn et al., 2005; Weinberger et al., 2001).

Several cognitive abnormalities associated with schizophrenia have consistently been found with increased prevalence in healthy siblings of patients with schizophrenia, including healthy monozygotic co-twins. The evidence from twin studies suggests that the cognitive deficits related to IQ, attention and working memory are particularly heritable and risk-associated traits (Cannon et al., 2000; Toulopoulou et al., 2007). Functional neuroimaging of prefrontal cortical changes during working memory and cognitive control also have been observed to be familial and heritable (Callicott et al., 2003a; Fusar-Poli et al., 2007; MacDonald, Pogue-Geile, Johnson, & Carter, 2003). Thus, the genetic mechanisms of these intermediate cognitive brain processes are therefore potentially tractable using these neuroimaging paradigms, and this approach is the focus for our review on the imaging genetics of prefrontal brain systems.

NEUROBIOLOGICAL AND COMPUTATIONAL MODELS OF WORKING MEMORY

Working memory has been studied extensively in animal and computational models (see also reviews by O'Reilly, 2006; Robbins & Roberts, 2007; Rolls, Loh, Deco, & Winterer, 2008). Seminal work on single-unit recordings in non-human primates have demonstrated that neurons located around the principal sulcus in the dorsolateral prefrontal cortex exhibit activity corresponding to maintaining information in an active state during the delay-period of the working memory paradigm (Goldman-Rakic, 1987, 1990). Numerous functional neuroimaging studies have since elaborated the key role of the prefrontal cortex in human working memory. It is also clear that the neural system supporting working memory engages other distant cortical areas. These include the posterior parietal cortex, the inferotemporal cortex, the cingulate gyrus and hippocampus – all of which have anatomical connections with the prefrontal cortex (Cavada & Goldman-Rakic, 1989; Fuster, 1997; Goldman-Rakic, 1988, 1999; Jonides et al., 1998; Petrides & Pandya, 1984; Selemon & Goldman-Rakic, 1985).

Neural mechanisms of working memory are critically dependent on dopaminergic modulation of glutamatergic and GABAergic brain systems during the processing of delay-period maintenance and manipulation of information. Experiments with dopamine D1 receptor agonists and antogonists modulate memory-dependent prefrontal neural firing, with optimal tuning of activity following an inverted-U curve across incremental D1 stimulation (Sawaguchi & Goldman-Rakic, 1991, 1994; Williams & Goldman-Rakic, 1995). These data support the hypothesis that locally sustained firing of PFC neurons crucial in the maintenance of relevant information during the delay period of working memory are stabilized against distracters through dopamine D1-receptors (Williams & Goldman-Rakic, 1995). These D1 receptors allow a focused augmentation of the task-relevant signal (Durstewitz, Seamans, & Sejnowski, 2000; Seamans, Gorelova, Durstewitz, & Yang, 2001b) by enhancing NMDA-receptor mediated post-synaptic currents in prefrontal pyramidal neurons, which are also active during the delay period (Seamans, Durstewitz, Christie, Stevens, & Sejnowski, 2001a; Wang, 1999; Wang & O'Donnell, 2001). Concurrently, D1-receptors also trigger a tonic increase in the firing of GABAergic inhibitory interneurons acting further afield, reducing irrelevant firing activity while allowing the focused increase in task-relevant activity, thus optimizing neural signal-to-noise (Seamans et al., 2001b).

On the other hand, dopamine D2-receptor signaling might play a crucial complementary role in working memory. D2 signals are hypothesized to enable new information to be rapidly updated and/or manipulated online, by transiently reducing the barriers for new information signals to be established in the cortical networks. This D2 process is key in ensuring that salient new information from cognitive computations of task-goal directed or reward information, is rapidly processed and updated online; this information processing network involves the PFC, posterior cortex, and striatum (Goldman-Rakic, 1995; Mink, 1996; O'Reilly, 2006; Tanaka et al., 2004). Thus, dopaminergic, glutamatergic and GABAergic systems are finely tuned to act in concert to maintain neural signal-to-noise critical for effective information processing.

IMAGING GENETICS AND DOPAMINERGIC MECHANISMS OF HUMAN PREFRONTAL CORTICAL FUNCTION

Dopaminergic, glutamatergic, other neurotransmitter systems, and related genes have been implicated in the prefrontal neuropathology of schizophrenia (Harrison & Weinberger, 2005). Given that the human prefrontal cortex is onto- and phylogenetically late developing, biologically complex and implicated in neuropsychiatric disease, understanding the mechanisms by which genetic susceptibility impacts active human prefrontal function is a clinically important challenge. Combining neuroimaging and genetics experiments could be a promising research strategy for elucidating such brain mechanisms (Egan et al., 2001a; Meyer-Lindenberg & Weinberger, 2006; Tan et al., 2008a; Weinberger et al., 2001). In what follows, we will elaborate on recent imaging strategies examining the effects of genetic variation on component brain systems engaged in working memory. We will also examine epistatic interactions of genes reflecting the non-additive activity-dependent cross-talk across various neurotransmitter systems (Tan et al., 2007c) and intracellular signaling mechanisms (Tan et al., 2008b), that has begun to translate basic neurobiology and genetic risk into human systems neuroscience.

Initial experiments on the impact of dopaminergic gene variation on cortical function examined catechol-O-methyltransferase (COMT). This is a major enzyme in prefrontal synaptic dopamine catabolism with a critical role in prefrontal cortical dopamine signaling because of the relative lack of dopamine transporters in this region (Chen et al., 2004; Lewis et al., 2001; Seasack, Hawrylak, Matus, Guido, & Levey, 1998). A common polymorphism in the COMT gene resulting from a valine-to-methionine Val(108/158)Met substitution gives rise to a significant reduction in its enzymatic activity (Chen et al., 2004; Lachman et al., 1996; Lotta et al., 1995). This was found to correspond to reduced prefrontal dopamine in proportion to the Val-allele load. Located on chromosome 22q11, COMT is also deleted in velocardiofacial syndrome, a condition that has 20 times increased risk for psychosis (Murphy, 2002). However, while this susceptibility locus has been implicated in some meta-analyses of linkage to schizophrenia (Badner & Gershon, 2002; Lewis et al., 2003), the effect on risk for schizophrenia of the specific COMT Val(108/158)Met polymorphism is small and inconsistent (Allen et al., 2008; Fan et al., 2005; Glatt, Faraone, & Tsuang, 2003). This is not surprising given the manifold factors associated with schizophrenia pathogenesis, such as the involvement of combinations of single nucleotide polymorphisms or haplotypes on the gene (Palmatier et al., 2004; Shifman et al., 2002), interactions across different susceptibility genes (Nicodemus et al., 2007), and interactions between genes and environment (Caspi et al., 2005).

In contrast, the effect of the COMT Val(108/158)Met polymorphism on intermediate measures of human brain function have reflected predictions from the basic cellular models of prefrontal dopamine described earlier. Reduced prefrontal dopamine in COMT Val-carriers should lead to decreased tonic D1-receptor activation. This might result firstly, in reduced cortical signal-to-noise; and secondly, in a relatively inefficient prefrontal cortical activation pattern if performance accuracy is still maintained; and ultimately, in reduced performance outcomes in working memory and executive function tasks. Each of these predictions have been repeatedly observed. COMT genotypes consistently account for a small but significant (about 3–4%) variance in performance on frontal-lobe tests, with poorer performance by subjects carrying the Val rather than the Met allele, even in subjects without neuropsychiatric disease (de Frias et al., 2006; Egan et al., 2001b; Goldberg et al., 2003; Malhotra et al., 2002; Nolan, Bilder, Lachman, & Volavka, 2004). Analogous results were also obtained in healthy children (Diamond, Biriand, Fossella, & Gehlbach, 2004). Correspondingly, using fMRI to study cortical activity, healthy COMT Val-allele carriers engaged relatively greater prefrontal cortical activation to perform the working memory task at the same speed and accuracy as those with the Met allele; this finding is consistent with the interpretation that Val carriers are relatively less efficient without advantages in performance accuracy or reaction time (Egan et al., 2001b; Mattay et al., 2003; Meyer-Lindenberg et al., 2006). As might be predicted by the greater dependence on dorsolateral prefrontal cortical processing for more complex executive tasks (D'Esposito, Postle, Ballard, & Lease, 1999; Deco & Rolls, 2005b; Deco, Rolls, & Horwitz, 2004; Koechlin, Ody, & Kouneiher, 2003), the COMT effects in dorsolateral prefrontal cortex were more prominent at higher working memory loads (Mattay et al., 2003). The study by Mattay et al. (2003) also demonstrated that the inefficient prefrontal cortical activation in Val-homozygotes could be improved by the dopamimetic amphetamine. In these less efficient Val-homozygous individuals, amphetamine resulted in a more focused reduction of dorsolateral prefrontal activation, presumably shifting these individuals closer to the peak on the inverted-U dopamine tuning curve (Mattay et al., 2003). Conversely, Met-homozygotes, already closer to peak dopaminergic tuning at baseline, became relatively inefficient after amphetamine, and thus appeared to have been shifted off the peak of the curve by amphetamine.

Cortical information processing, especially in prefrontal cortex, impacts on the regulation of dopamine activity in the mesencephalon (Winterer & Weinberger, 2004). This is important in learning as reward signals emanating from brainstem dopamine neuronal firing should correspond to prefrontal cortical executive action for learning to take place. COMT, presumably via its actions at the cortical level, appears to impact on dopamine activity in the brainstem. In a study of normal postmortem human brainstem, individuals with a COMT Val/Val genotype had twice the expression of the mRNA for tyrosine hydroxylase, the rate limiting biosynthetic enzyme for dopamine (Akil et al., 2003). Remarkably, this relationship has been confirmed in a PET imaging study of normal living subjects. COMT Val-carriers were found to have relatively increased midbrain dopamine synthesis (measured with f-18 flurodopa uptake) that correlated negatively with N-back dorsolateral prefrontal cortical activation (measured with O-15 H2O regional cerebral blood flow), while prefrontal activation in Met-homozygotes positively correlated with midbrain dopamine synthesis (Meyer-Lindenberg et al., 2005). These reciprocal relationships fit tightly to the inverted-U dopamine tuning curve, whereby increased midbrain dopamine synthesis in COMT-Val carriers with lower prefrontal dopamine tended to restore efficient cortical activation, while COMT-Met homozygotes became less efficient with increased dopamine synthesis. These multi-modality PET imaging findings support conceptualizations of dopamine's role in the fine tuning of neural circuits (Akil et al., 2003; Durstewitz et al., 2000; Seamans & Yang, 2004), as do the fMRI data on amphetamine and COMT (Mattay et al., 2003).

Less is known, however, about how these functional dopaminergic effects correspond spatially to the differing prefrontal-parietal-striatal networks during specific human working memory sub-processes. It has been conceptualized that the lateral prefrontal cortex is hierarchically organized, where more dorsal and anterior prefrontal regions (e.g., dorsolateral PFC, DLPFC: Brodmann Areas BA 9, 10 and 46) engaged higher-order processing such as in manipulating information or applying them in context, while the ventrolateral regions (e.g., VLPFC: BAs 44, 45 and 47) were engaged during simpler operations (D'Esposito et al., 1999; Deco & Rolls, 2005a; Fuster, 1997; Koechlin et al., 2003; Sakai & Passingham, 2002). If indeed dopamine tuning of cortical neural assemblies is critical for their effective function in working memory (Seamans & Yang, 2004; Vijayraghavan, Wang, Birnbaum, Williams, & Arnsten, 2007; Williams & Goldman-Rakic, 1995), it might be expected that these dopamine effects would also be observed in these specific, hierarchically organized PFC functions. Moreover, if dopamine function is especially implicated in the updating and stabilization of representations (O'Reilly, 2006; Seamans & Yang, 2004; Tanaka et al., 2004), then executive working memory tasks emphasizing encoding, manipulating, and temporally integrating information should be disproportionately more dependent on changes in cortical dopaminergic signaling than tasks emphasizing simple retrieval of already stabilized information. Some of the former processes are also likely to involve the dopamine-rich striatum, which has been postulated to have intimate connections to cortex in implementing the selective gating of information during rapid updating and manipulation in working memory (Alexander, DeLong, & Strick, 1986; Goldman-Rakic, 1995; Gruber, Dayan, Gutkin, & Solla, 2006; O'Reilly & Frank, 2006). For example, manipulating numerical representations should engage dopamine-dependent processes in the striatum, prefrontal cortex, and number-sensitive regions in the parietal cortex (Dehaene, Piazza, Pinel, & Cohen, 2003; Hubbard, Piazza, Pinel, & Dehaene, 2005). On the other hand, anterior regions in the DLPFC might be more specifically engaged during dopamine-dependent processing of higher-order temporal or episodic aspects of working memory (Koechlin et al., 2003; Sakai & Passingham, 2002).

Indeed, it has been observed using genetic imaging of COMT that dopaminergic modulation integral to differing levels of working memory processing occurred with a degree of spatial and process specificity within the human prefrontal-parietal-striatal network (Tan et al., 2007b). In an event-related fMRI task dissociating component numerical working memory sub-processes, baseline numerical size comparison engaged ventrolateral prefrontal cortical activation that correlated with COMT Val-allele load (COMT Val > Met), while further performance of arithmetic transformations engaged this genotype effect in dorsolateral prefrontal cortex (DLPFC), as well as in parietal and striatal regions. Critically, additional temporal integration of information in working memory disproportionately engaged greater COMT Val > Met effects only at DLPFC. COMT Val > Met effects were also observed in DLPFC during encoding of new information into working memory, but not at its subsequent retrieval. Thus, temporal updating operations, but less so the retrieval of already encoded representations, engaged relatively specific dopaminergic tuning at the DLPFC. Manipulating and rapidly updating representations were sensitive to dopaminergic modulation of neural signaling in a larger prefrontal-parietal-striatal network. The relatively specific engagement of prefrontal-parietal-striatal dopaminergic modulation during these computational tasks support their role in the effective control of rapid switching and stabilization processes intrinsic in such tasks that engage the manipulation of information. This is also consistent with models predicting basal ganglia coupling of prefrontal cortex and modality-specific (e.g., numerical) regions in the posterior cortex, in order to effect this highly selective information transformation and updating; these models also propose that dopamine is critical in the implementation of these targeted gating processes in the human brain (Gruber et al., 2006; O'Reilly & Frank, 2006).

On the other hand, processes involved in the manipulation of information might be distinguished from those engaged in the temporal integration of information in WM. The latter were associated with more prominent dopaminergic modulation within the anterior DLPFC rather than in the striatum or posterior parietal cortex. This observation argues that dopaminergic processes in these DLPFC regions might more critically mediate higher-order temporal processes, such as when contextual information is encoded for future operations, or when new probe information needs to be integrated with that encoded previously. Taken together with propositions that these higher-order processes could engage more overall inhibitory (Deco & Rolls, 2005a) or biasing (Miller & Cohen, 2001) cognitive control putatively engaging greater dopamine D1 than D2 mechanisms (Durstewitz et al., 2000; Seamans & Yang, 2004), the former postulated to predominate in the PFC (Goldman-Rakic, Lidow, & Gallager, 1990), one might speculate that our systems-level findings at these DLPFC regions during live human cognition could reflect greater D1 dopaminergic modulation during higher-order temporal integration of information. Conversely, rapid updating in manipulation involving the

DLPFC, striatum and posterior cortex might reflect the engagement of predominantly D2 mechanisms (Goldman-Rakic, 1995; Gruber et al., 2006; Mink, 1996; O'Reilly & Frank, 2006; Seamans & Yang, 2004). In an analogous cognitive model, three dopamine system genes, DARPP-32, DRD2 and COMT, have been shown to impact differentially on specific processes in prefrontal-striatally mediated reinforcement learning from positive and negative outcomes (Frank, Moustafa, Haughey, Curran, & Hutchison, 2007).

EPISTATIC GENE MECHANISMS OF HUMAN PREFRONTAL CORTICAL FUNCTION

Glutamatergic abnormalities, in addition to dopamine, are important in schizophrenia and working memory function. The NMDA receptor system is a critical partner in working memory processes (Jackson, Homayoun, & Moghaddam, 2004; Seamans & Yang, 2004; Wang & O'Donnell, 2001), and disease-related changes in glutamate signaling could well impair working memory. For example, the metabotropic glutamate receptor, GRM3 on chromosome 7q21-22, modulates NMDA receptor transmission (Egan et al., 2004; Harrison & Weinberger, 2005; Moghaddam & Adams, 1998). GRM3 regulates synaptic glutamate via a presynaptic mechanism and by regulating the expression of the glial glutamate transporter, which inactivates synaptic glutamate. A polymorphism in intron 2 and related haplotypes were significantly associated with schizophrenia in several samples (Chen et al., 2005; Egan et al., 2004; Fallin et al., 2005; Martí, Cichon, Propping, & Nöthen, 2002), though negative studies also have been reported (Norton et al., 2005). Risk variants in GRM3 may also influence alternative splicing of GRM3 mRNA and its products (Sartorius et al., 2006). In postmortem brain, the risk allele is associated with reduced prefrontal glial glutamate transporter EAAT2, a protein modulating synaptic glutamate (Egan et al., 2004). Consistent with the role of the glutamatergic system in schizophrenia and working memory, the risk allele was associated with inefficient prefrontal cortical fMRI activation and reduced working memory performance even in normal subjects (Egan et al., 2004).

Importantly, given the tight relationships governing dopaminergic and glutamatergic (and GABAergic) dynamics in the biology of working memory (Durstewitz & Seamans, 2002; Lisman, Fellous, & Wang, 1998; Sawaguchi & Goldman-Rakic, 1991; Seamans & Yang, 2004; Wang, 1999; Williams & Goldman-Rakic, 1995), and their putatively greater involvement in executive aspects of working memory at the dorsolateral prefrontal cortex (D'Esposito et al., 1999; Deco & Rolls, 2005b; Deco et al., 2004; Koechlin et al., 2003; Mattay et al., 2003; Tan et al., 2007b), we would expect that higher-order working memory processes taxing dorsolateral prefrontal

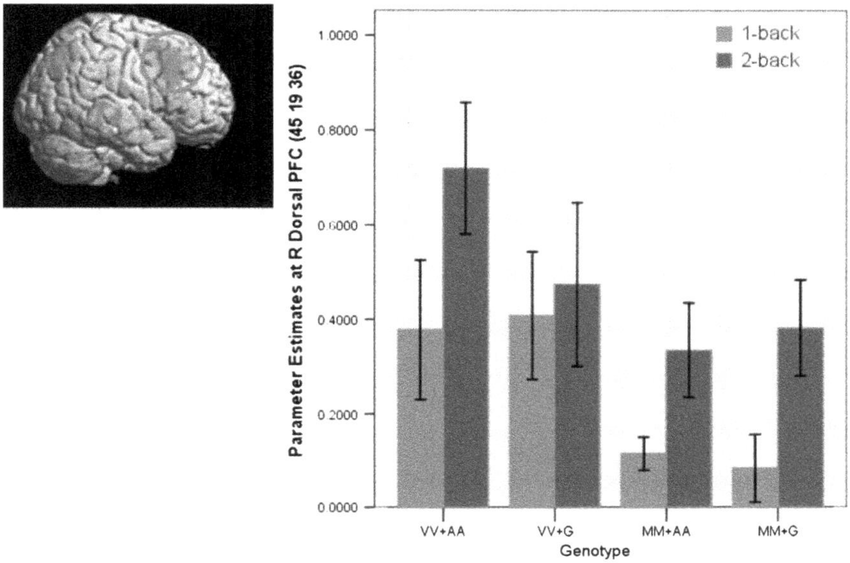

Figure 1. Epistatic interaction between COMT and GRM3 on prefrontal brain function. Higher load working memory processes engaging dorsolateral prefrontal cortex was disproportionately inefficient in the context of combined suboptimal COMT and GRM3 risk alleles $F(1, 25) = 4.47$, $p = .045$; adapted from Tan et al. (2007).

cortex might be more vulnerable to the combined effect of suboptimal dopaminergic and glutamatergic influence. Consistent with the interplay of cortical macrocircuits suggested by these possibilities, a recent fMRI study revealed that the integrity of higher executive areas in the dorsolateral prefrontal cortex could be disproportionately compromised and inefficient in the presence of combined deleterious COMT and GRM3 genotypes in normal subjects (Figure 1) (Tan et al., 2007c). Thus, genetic variation impacting important nodes in the dopamine and glutamatergic systems at a molecular level, when combined, had disproportionate or non-additive influence on executive cognitive brain function at the human systems level. If so, we might also expect key nodes at the intracellular signaling cascades of these receptor systems to play similarly important roles, as we extend our search for molecules impacting variation in human brain function.

PREFRONTAL FUNCTIONAL GENETIC LINKS WITH INTRACELLULAR SIGNALING MOLECULES

Classically, D1 receptors, implicated in the maintenance of relevant information during the working memory delay period (Williams & Goldman-Rakic,

1995), couple through $G\alpha_s$ to stimulate the production of cyclic adenosine monophosphate (cAMP) and the activity of protein kinase A (PKA) (Greengard, 2001). Conversely, D2 receptors, which in neural models play critical roles marking salience, prediction errors, and updating and manipulating new information (O'Reilly, 2006), couple through $G\alpha_{i/o}$ to reduce cAMP production and PKA activity (Greengard, 2001). Downstream from PKA, "dopamine- and cAMP-regulated phosphoprotein of molecular weight 32" (DARPP-32) is a key signaling integrator that regulates an array of subsequent neurophysiological processes (Greengard, 2001). Indeed, human genetic variation in *DARPP-32* has been found to impact normal human variation in frontostriatal cognitive performance, in neostriatal volume and physiologic activation and functional connectivity between the striatum and prefrontal cortex, as well as risk for schizophrenia (Meyer-Lindenberg et al., 2007).

In addition to the cAMP-PKA pathway, D2 receptors may also signal through an AKT1 (protein kinase B)-GSK-3 signaling cascade via β-arrestin 2 (Beaulieu et al., 2005). Of note, this AKT-GSK-3 pathway influences the expression of dopamine-associated psychomotor behaviors that, in transgenic mice models, have been predictably modulated by dopaminergic drugs; this pathway also appears independent of the cAMP-associated one, and represents a novel means by which D2 receptor signaling and associated cognitive and neuropsychiatric effects could be mediated (Beaulieu et al., 2004, 2005; Lai et al., 2006). *AKT1* knockout mice, in particular, evidenced abnormal prepulse inhibition of startle (Emamian, Hall, Birnbarum, Karayiogou, & Gogos, 2004) and poorer working memory performance under dopaminergic agonist challenge, as well as concurrent changes in prefrontal pyramidal dendritic ultrastructure, possibly mediated by downstream alterations in the expression of genes controlling neuronal development in prefrontal cortex (Lai et al., 2006).

We have then taken a strategy to examine genetic association of *AKT1* with human brain phenotypes related to dopamine function (Tan et al., 2008b). In examining the effect of a genetic variant in *AKT1* that consistently affected the reduced expression of AKT1 protein levels (Harris et al., 2005; Tan et al., 2008b), we found that this same SNP influenced, even in healthy individuals, frontostriatal cognitive tasks taxing processing speed, IQ and executive cognitive control, cardinal DA-mediated cognitive functions (Alexander et al., 1986; Kane & Engle, 2002; Pantelis et al., 1997; Salthouse, 1996), as well as "tuning" of the prefrontal physiological phenotype previously linked to cortical dopamine function during working memory (Egan et al., 2001b; Mattay et al., 2002, 2003; Tan et al., 2007b). The same genotype also predicted reduced gray matter volume in parts of the frontostriatal network (Tan et al., 2008b). As a further test that the AKT1 effects were linked to dopaminergic function, we found that the same prefrontal regions showing *AKT1* main effects in fMRI and MRI volumetry evidenced epistasis with the *COMT* Val/Met (Tan et al., 2008b). Individuals with COMT-Val homozygous genetic background had exaggerated (i.e., non-additive) *AKT1* effects with "inefficient" activation in replicated datasets at the prefrontal cortex (Figure 2). In terms of brain structure, gray matter volume from the prefrontal cortex also showed the *AKT1*-by-*COMT* interaction in that individuals with combined deleterious *AKT1*

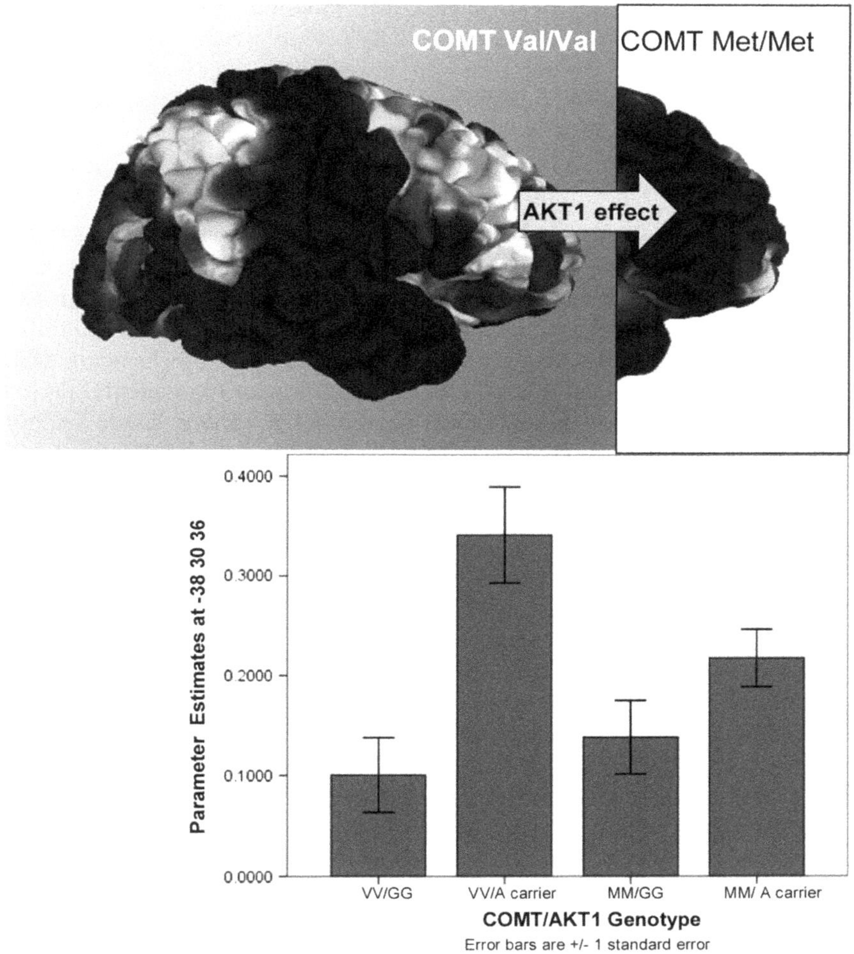

Figure 2. Epistatic interaction between AKT1 and COMT. Here, individuals with the AKT1 allele associated with reduced gene expression showed disproportionately inefficient dorsolateral prefrontal cortical activity in the background of relatively deleterious COMT Val allele, $F(1, 42) = 4.466, p = .041$; adapted from Tan et al. (2008).

minor and *COMT* Val-alleles had disproportionately reduced gray matter volume. Thus, these results provided multiple lines of converging evidence implicating *AKT1* gene effects that influenced protein expression as well as systems-level human prefrontal structure and function. The results were consistent with pre-clinical evidence that coupled AKT1 to dopaminergic signaling and downstream effects on prefrontal cellular structure and cognition (Beaulieu et al., 2005; Emamian et al., 2004; Lai et al., 2006; Wei et al., 2007), and suggested that these brain mechanisms impacted the biology of active human cognitive function. The data also suggest that the mechanisms of prior associations of *AKT1* with psychosis, a condition associated with DA abnormalities in brain (Laruelle, Kegeles, & Abi-Dargham, 2003; Winterer & Weinberger, 2004) and treated with antidopaminergic drugs, could involve these biologic processes.

CONCLUSION

In this review, we have highlighted a selection of recent findings through which heritable human neuroimaging intermediate phenotypes could provide a window to examine genetic mechanisms of active prefrontal cognitive processing relevant to schizophrenia. Genetic variation influencing task-related prefrontal cortical function was consistent with fundamental predictions based on the biology of cortical dopaminergic tuning. These findings also extended the basic biological data to implicate molecules impacting variation in active human brain function, potentially mirroring component disease-related brain processes. The findings of interacting genetic elements consistent with the cross-talk across neurotransmitter systems and their intracellular signaling pathways arguably contributed another level of empirical validation to the strategy to identify molecules whose genetic variation could be of significant combined influence on human brain function at the network systems level. Indeed, an increasing number of recent pharmacological and gene-based human studies (Apud et al., 2007; Bertolino et al., 2004, 2006, 2009; Gothelf et al., 2005; McIntosh et al., 2007), including in patients with Parkinson's disease (Mattay et al., 2002; Williams-Gray et al., 2007) have also been consistent with the main findings of dopamine-dependent gene effects on prefrontal function highlighted in this review. Recent imaging genetics work have also examined the serotonergic system (Hariri et al., 2002; Heinz et al., 2005) and epistatic interactions therein (Pezawas et al., 2008) of relevance to mood and anxiety disorders.

If confirmed and extended, imaging genetics could be an attractive strategy to discover key molecules influencing active human brain function impacting neuropsychiatric disease pathophysiology and potential treatment. It is early days yet to explore the utility of these data on potential treatment development, but encouraging parallels might be drawn from recent independent data suggesting that novel treatments targeting metabotropic glutamate receptors were potentially beneficial in treating symptoms of schizophrenia (Patil et al., 2007). It is an open question as to whether treatments targeting, for example GRM3 or AKT1 could impact cognitive brain processes such as working memory, perhaps in concert with dopaminergic modulation and as a function of individual genotype status.

REFERENCES

Akil, M., Kolachana, B. S., Rothmond, D. A., Hyde, T. M., Weinberger, D. R., & Kleinman, J. E. (2003). Catechol-O-methyltransferase genotype and dopamine regulation in the human brain. *Journal of Neuroscience, 23*, 2008–2013.

Alexander, G. E., DeLong, M. R., & Strick, P. L. (1986). Parallel organization of functionally segregated circuits linking basal ganglia and cortex. *Annual Review of Neuroscience, 9*, 357–381.

Allen, N. C., Bagade, S., McQueen, M. B., et al. (2008). Systematic meta-analyses and field synopsis of genetic association studies in schizophrenia: The SzGene database. *Nature Genetics, 40*, 827.

Apud, J. A., Mattay, V., Chen, J., et al. (2007). Tolcapone improves cognition and cortical information processing in normal human subjects. *Neuropsychopharmacology, 32*, 1011–1020.

Baddeley, A. D. (2003). Working memory: Looking back and looking forward. *Nature Reviews Neuroscience, 4*, 829–839.

Badner, J. A., & Gershon, E. S. (2002). Meta-analysis of whole-genome linkage scans of bipolar disorder and schizophrenia. *Molecular Psychiatry, 7*, 405–411.

Beaulieu, J. M., Sotnikova, T. D., Marion, S., Lefkowitz, R. J., Gainetdinov, R. R., & Caron, M. G. (2005). An Akt/β-Arrestin 2/PP2A signaling complex mediates dopaminergic neurotransmission and behavior. *Cell, 122*, 261–273.

Beaulieu, J. M., Sotnikova, T. D., Yao, W. D., et al. (2004). Lithium antagonizes dopamine-dependent behaviors mediated by an AKT/glycogen synthase kinase 3 signaling cascade. *Proceedings of the National Academy of Sciences of the United States of America, 101*, 5099–5104.

Bertolino, A., Blasi, G., Latorre, V., et al. (2006). Additive effects of genetic variation in dopamine regulating genes on working memory cortical activity in human brain. *Journal of Neuroscience, 26*, 3918–3922.

Bertolino, A., Caforio, G., Blasi, G., et al. (2004). Interaction of COMT Val108/158 Met genotype and Olanzapine treatment on prefrontal cortical cunction in patients with schizophrenia. *American Journal of Psychiatry, 161*, 1798–1805.

Bertolino, A., Fazio, L., Caforio, G., et al. (2009). Functional variants of the dopamine receptor D2 gene modulate prefronto-striatal phenotypes in schizophrenia. *Brain, 132*(2), 417–425.

Blackwood, D. H., St Clair, D. M., Muir, W. J., & Duffy, J. C. (1991). Auditory P300 and eye tracking dysfunction in schizophrenic pedigrees. *Archives of General Psychiatry, 48*, 899–909.

Callicott, J. H., & Weinberger, D. R. (1999). Neuropsychiatric dynamics: The study of mental illness using functional magnetic resonance imaging. *European Journal of Radiology, 30*, 95.

Callicott, J. H., Mattay, V. S., Bertolino, A., et al. (1999). Physiological characteristics of capacity constraints in working memory as revealed by functional MRI. *Cerebral Cortex, 9*, 20–26.

Callicott, J. H., Bertolino, A., Mattay, V. S., et al. (2000). Physiological dysfunction of the dorsolateral prefrontal cortex in schizophrenia revisited. *Cerebral Cortex, 10*, 1078–1092.

Callicott, J. H., Egan, M. F., Mattay, V. S., et al. (2003a). Abnormal fMRI response of the dorsolateral prefrontal cortex in cognitively intact siblings of patients with schizophrenia. *American Journal of Psychiatry, 160*, 709–719.

Callicott, J. H., Mattay, V. S., Verchinski, B. A., Marenco, S., Egan, M. F., & Weinberger, D. R. (2003b). Complexity of prefrontal cortical dysfunction in schizophrenia: More than up or down. *American Journal of Psychiatry, 160*, 2209–2215.

Cannon, T. D., Huttunen, M. O., Lonnqvist, J., et al. (2000). The inheritance of neuropsychological dysfunction in twins discordant for schizophrenia. *American Journal of Human Genetics, 67*, 369–382.

Caspi, A., Moffitt, T. E., Cannon, M., et al. (2005). Moderation of the effect of adolescent-onset cannabis use on adult psychosis by a functional polymorphism in the catechol-O-methyl-transferase gene: Longitudinal evidence of a gene X environment interaction. *Biological Psychiatry, 57*, 1117–1127.

Cavada, C., & Goldman-Rakic, P. S. (1989). Posterior parietal cortex in rhesus monkey: II. Evidence for segregated corticocortical networks linking sensory and limbic areas with the frontal lobe. *The Journal of Comparative Neurology, 287*, 422–445.

Chen, J., Lipska, B. K., Halim, N., et al. (2004). Functional analysis of genetic variation in catechol-O-methyltransferase (COMT): Effects on mRNA, protein, and enzyme activity in postmortem human brain. *American Journal of Human Genetics, 75*, 807–821.

Chen, Q., He, G., Chen, Q., et al. (2005). A case-control study of the relationship between the metabotropic glutamate receptor 3 gene and schizophrenia in the Chinese population. *Schizophrenia Research, 73*, 21.

D'Esposito, M., Postle, B. R., Ballard, D., & Lease, J. (1999). Maintenance versus manipulation of information held in working memory: An event-related fMRI study. *Brain and Cognition, 41*, 66–86.

de Frias, C. M., Annerbrink, K., Westberg, L., Eriksson, E., Adolfsson, R., & Nilsson, L. G. (2006). Catechol O-methyltransferase Val158Met polymorphism is associated with cognitive performance in nondemented adults. *Journal of Cognitive Neuroscience, 17*, 1018–1025.

Deco, G., & Rolls, E. T. (2005a). Attention, short-term memory, and action selection: A unifying theory. *Progress in Neurobiology, 76*, 236–256.

Deco, G., & Rolls, E. T. (2005b). Sequential memory: A putative neural and synaptic dynamical mechanism. *Journal of Cognitive Neuroscience, 17*, 294–307.

Deco, G., Rolls, E. T., & Horwitz, B. (2004). "What" and "where" in visual working memory: A computational neurodynamical perspective for integrating fMRI and single-neuron data. *Journal of Cognitive Neuroscience, 16*, 683–701.

Dehaene, S., Piazza, M., Pinel, P., & Cohen, L. (2003). Three parietal circuits for number processing. *Cognitive Neuropsychology, 20*, 487–506.

Diamond, A., Biriand, L., Fossella, J., & Gehlbach, L. (2004). Genetic and neurochemical modulation of prefrontal cognitive functions in children. *American Journal of Psychiatry, 161*, 125–132.

Durstewitz, D., & Seamans, J. K. (2002). The computational role of dopamine D1 receptors in working memory. *Neural Networks, 15*, 561–572.

Durstewitz, D., Seamans, J. K., & Sejnowski, T. J. (2000). Neurocomputational models of working memory. *Nature Neuroscience, 3*, 1184–1191.

Egan, M. F., Goldberg, T. E., Gscheidle, T., et al. (2001a). Relative risk for cognitive impairments in siblings of patients with schizophrenia. *Biological Psychiatry, 50*, 98–107.

Egan, M. F., Goldberg, T. E., Kolachana, B. S., et al. (2001b). Effect of COMT Val108/158 Met genotype on frontal lobe function and risk for schizophrenia. *Proceedings of the National Academy of Sciences of the United States of America, 98*, 6917–22.

Egan, M. F., Straub, R. E., Goldberg, T. E., et al. (2004). Variation in GRM3 affects cognition, prefrontal glutamate, and risk for schizophrenia. *Proceedings of the National Academy of Sciences of the United States of America*, *101*, 12604–12609.

Emamian, E. S., Hall, D., Birnbarum, M. J., Karayiogou, M., & Gogos, J. A. (2004). Convergent evidence for impaired AKT1-GSK3b signaling in schizophrenia. *Nature Genetics*, *36*, 131–137.

Fallin, M. D., Lasseter, V. K., Avramopoulos, D., et al. (2005). Bipolar I disorder and schizophrenia: A 440-single-nucleotide polymorphism screen of 64 candidate genes among Ashkenazi Jewish case-parent trios. *American Journal of Human Genetics*, *77*, 918–936.

Fan, J. B., Zhang, C. S., Gu, N. F., et al. (2005). Catechol-O-methyltransferase gene Val/Met functional polymorphism and risk of schizophrenia: A large-scale association study plus meta-analysis. *Biological Psychiatry*, *57*, 139–144.

Frank, M. J., Moustafa, A. A., Haughey, H. M., Curran, T., & Hutchison, K. E. (2007). Genetic triple dissociation reveals multiple roles for dopamine in reinforcement learning. *Proceedings of the National Academy of Sciences*, *104*, 16311–16316.

Fusar-Poli, P., Perez, J., Broome, M., et al. (2007). Neurofunctional correlates of vulnerability to psychosis: A systematic review and meta-analysis. *Neuroscience & Biobehavioral Reviews*, *31*, 465.

Fuster, J. M. (1997). *The prefrontal cortex – anatomy, physiology, and neuropsychology of the frontal lobe*. Philadelphia: Lippincott-Raven.

Glahn, D. C., Ragland, J. D., Abramoff, A., et al. (2005). Beyond hypofrontality: A quantitative meta-analysis of functional neuroimaging studies of working memory in schizophrenia. *Human Brain Mapping*, *25*, 60–69.

Glatt, S. J., Faraone, S. V., & Tsuang, M. T. (2003). Association between a functional catechol-O-methyltransferase gene polymorphism and schizophrenia: Meta-analysis of case-control and family-based studies. *American Journal of Psychiatry*, *160*, 469–476.

Goldberg, T. E., & Weinberger, D. R. (1996). Effects of neuroleptic medications on the cognition of patients with schizophrenia: A review of recent studies. *Journal of Clinical Psychiatry*, *57*, 62–65.

Goldberg, T. E., Weinberger, D. R., Berman, K. F., Pliskin, N. H., & Podd, M. H. (1987). Further evidence for dementia of the prefrontal type in schizophrenia? A controlled study of teaching the Wisconsin Card Sorting Test. *Archives of General Psychiatry*, *44*, 1008–1014.

Goldberg, T. E., Torrey, E. F., Gold, J. M., et al. (1995). Genetic risk of neuropsychological impairment in schizophrenia: A study of monozygotic twins discordant and concordant for the disorder. *Schizophrenia Research*, *17*, 77.

Goldberg, T., Egan, M., Gscheidle, T., et al. (2003). Executive subprocesses in working memory: Relationship to catechol-O-methyltransferase Val158Met genotype and schizophrenia. *Archives of General Psychiatry*, *60*, 889–896.

Goldman, A. L., Pezawas, L., Mattay, V. S., et al. (2008). Heritability of brain morphology related to schizophrenia: A large-scale automated magnetic resonance imaging segmentation study. *Biological Psychiatry*, *63*(5), 475–483.

Goldman-Rakic, P. S. (1987). Circuitry of primate prefrontal cortex and regulation of behavior by representational memory. In F. Plum & V. Mountcastle (Eds.), *Handbook of physiology: The nervous system V* (Vol. 5, pp. 373–417). Bethesda, MD: American Physiological Society.

Goldman-Rakic, P. S. (1988). Topography of cognition: Parallel distributed networks in primate association cortex. *Annual Review of Neuroscience*, *11*, 137–156.

Goldman-Rakic, P. S. (1990). Cellular and circuit basis of working memory in prefrontal cortex of nonhuman primates. *Progress in Brain Research*, *85*, 325–336.

Goldman-Rakic, P. S. (1994). Working memory dysfunction in schizophrenia. *Journal of Neuropsychiatry and Clinical Neuroscience*, *6*, 348–357.

Goldman-Rakic, P. S. (1995). Cellular basis of working memory. *Neuron*, *14*, 477.

Goldman-Rakic, P. S. (1999). The physiological approach: Functional architecture of working memory and disordered cognition in schizophrenia. *Biological Psychiatry*, *46*, 650–661.

Goldman-Rakic, P. S., Lidow, M. S., & Gallager, D. W. (1990). Overlap of dopaminergic, adrenergic, and serotoninergic receptors and complementarity of their subtypes in primate prefrontal cortex. *Journal of Neuroscience, 10*, 2125–2138.

Gothelf, D., Eliez, S., Thompson, T., et al. (2005). COMT genotype predicts longitudinal cognitive decline and psychosis in 22q11.2 deletion syndrome. *Nature Neuroscience, 8*, 1500–1502.

Gottesman, I. I., & Shields, J. (1972). *Schizophrenia and genetics. A twin study.* Vantage Point, New York: Academic Press.

Green, M. F. (1996). What are the functional consequences of neurocognitive deficits in schizophrenia? *American Journal of Psychiatry, 153*, 321–330.

Greengard, P. (2001). The neurobiology of slow synaptic transmission. *Science, 294*, 1024–1030.

Gruber, A. J., Dayan, P., Gutkin, B. S., & Solla, S. A. (2006). Dopamine modulation in the basal ganglia locks the gate to working memory. *Journal of Computational Neuroscience, 20*, 153–166.

Hariri, A. R., Mattay, V. S., Tessitore, A., et al. (2002). Serotonin transporter genetic variation and the response of the human amygdala. *Science, 297*, 400–403.

Harris, S. L., Gil, G., Robins, H., et al. (2005). Detection of functional single-nucleotide polymorphisms that affect apoptosis. *Proceedings of the National Academy of Sciences of the United States of America, 102*, 16297–16302.

Harrison, P. J., & Weinberger, D. R. (2005). Schizophrenia genes, gene expression, and neuropathology: On the matter of their convergence. *Molecular Psychiatry, 10*, 40–68.

Heinz, A., Braus, D. F., Smolka, M. N., et al. (2005). Amygdala-prefrontal coupling depends on a genetic variation of the serotonin transporter. *Nature Neuroscience, 8*, 20.

Hubbard, E. M., Piazza, M., Pinel, P., & Dehaene, S. (2005). Interactions between number and space in parietal cortex. *Nature Reviews Neuroscience, 6*, 435–448.

Ioannidis, J. P. A. (2005). Why most published research findings are false. *PLoS Medicine, 2*, e124.

Jackson, M. E., Homayoun, H., & Moghaddam, B. (2004). NMDA receptor hypofunction produces concomitant firing rate potentiation and burst activity reduction in the prefrontal cortex. *Proceedings of the National Academy of Sciences of the United States of America, 101*, 8467–8472.

Jones, P. B., Barnes, T. R. E., Davies, L., et al. (2006). Randomized controlled trial of the effect on quality of life of second- vs first-generation antipsychotic drugs in schizophrenia: Cost utility of the latest antipsychotic drugs in schizophrenia study (CUtLASS 1). *Archives of General Psychiatry, 63*, 1079–1087.

Jonides, J., Schumacher, E. H., Smith, E. E., et al. (1998). The role of parietal cortex in verbal working memory. *Journal of Neuroscience, 18*, 5026–5034.

Kane, M. J., & Engle, R. W. (2002). The role of prefrontal cortex in working-memory capacity, executive attention, and general fluid intelligence: An individual-differences perspective. *Psychonomic Bulletin & Review, 9*, 637.

Kety, S. S., Rosenthal, D., Wender, P. H., & Schulsinger, F. (1976). Studies based on a total sample of adopted individuals and their relatives: Why they were necessary, what they demonstrated and failed to demonstrate. *Schizophrenia Bulletin, 2*, 413–428.

Koechlin, E., Ody, C., & Kouneiher, F. T. (2003). The architecture of cognitive control in the human prefrontal cortex. *Science, 302*, 1181–1185.

Lachman, H. M., Papolos, D. F., Saito, T., Yu, Y. M., Szumlanski, C. L., & Weinshilboum, R. M. (1996). Human catechol-O-methyltransferase pharmacogenetics: Description of a functional polymorphism and its potential application to neuropsychiatric disorders. *Pharmacogenetics, 6*, 243.

Lai, W. S., Xu, B., Westphal, K. G. C., et al. (2006). Akt1 deficiency affects neuronal morphology and predisposes to abnormalities in prefrontal cortex functioning. *Proceedings of the National Academy of Sciences of the United States of America, 103*, 16906–16911.

Laruelle, M., Kegeles, L. S., & Abi-Dargham, A. (2003). Glutamate, dopamine, and schizophrenia: From pathophysiology to treatment. *Annals of the New York Academy of Science, 1003*, 138–158.

Lewis, C. M., Levinson, D. F., Wise, L. H., et al. (2003). Genome scan meta-analysis of schizophrenia and bipolar disorder, part II: Schizophrenia. *American Journal of Human Genetics, 73*, 34–48.

Lewis, D. A., Melchitzky, D. S., Seasack, S. R., Whitehead, R. E., Sungyoung, A. U. H., & Sampson, A. (2001). Dopamine transporter immunoreactivity in monkey cerebral cortex. *Journal of Comparative Neurology, 432*, 119–136.

Lieberman, J. A., Stroup, T. S., McEvoy, J. P., et al. (2005). Effectiveness of antipsychotic drugs in patients with chronic schizophrenia. *New England Journal of Medicine, 353*, 1209–1223.

Lisman, J. E., Fellous, J. M., & Wang, X. J. (1998). A role for NMDA-receptor channels in working memory. *Nature Neuroscience, 1*, 273–275.

Lotta, T., Vidgren, J., Tilgmann, C., et al. (1995). Kinetics of human soluble and membrane-bound catechol-O-methyltransferase: A revised mechanism and description of the thermolabile variant of the enzyme. *Biochemistry, 34*, 4202.

MacDonald, A. W., Pogue-Geile, M. F., Johnson, M. K., & Carter, C. S. (2003). A specific deficit in context processing in the unaffected siblings of patients with schizophrenia. *Archives of General Psychiatry, 60*, 57–65.

Malhotra, A. K., Kestler, L. J., Mazzanti, C., Bates, J. A., Goldberg, T., & Goldman, D. (2002). A functional polymorphism in the COMT gene and performance on a test of prefrontal cognition. *American Journal of Psychiatry, 159*, 652–654.

Manoach, D. S. (2003). Prefrontal cortex dysfunction during working memory performance in schizophrenia: Reconciling discrepant findings. *Schizophrenia Research, 60*, 285–298.

Martí, S. B., Cichon, S., Propping, P., & Nöthen, M. (2002). Metabotropic glutamate receptor 3 (GRM3) gene variation is not associated with schizophrenia or bipolar affective disorder in the German population. *American Journal of Medical Genetics, 114*, 46–50.

Mattay, V. S., Goldberg, T. E., Fera, F., et al. (2003). Catechol-O-methyltransferase val158-met genotype and individual variation in the brain response to amphetamine. *Proceedings of the National Academy of Sciences of the United States of America, 100*, 6186–6191.

Mattay, V. S., Tessitore, A., Callicott, J. H., et al. (2002). Dopaminergic modulation of cortical function in patients with Parkinson's disease. *Annals of Neurology, 51*, 156–164.

McIntosh, A. M., Baig, B. J., Hall, J., et al. (2007). Relationship of catechol-O-methyltransferase variants to brain structure and function in a population at high risk of psychosis. *Biological Psychiatry, 61*, 1127.

Meyer-Lindenberg, A. S., & Weinberger, D. R. (2006). Intermediate phenotypes and genetic mechanisms of psychiatric disorders. *Nature Reviews Neuroscience, 7*, 818–27.

Meyer-Lindenberg, A., Kohn, P. D., Kolachana, B., et al. (2005). Midbrain dopamine and prefrontal function in humans: Interaction and modulation by COMT genotype. *Nature Neuroscience, 8*, 594–596.

Meyer-Lindenberg, A. S., Nichols, T., Callicott, J. H., et al. (2006). Impact of complex genetic variation in COMT on human brain function. *Molecular Psychiatry, 11*, 867–877.

Meyer-Lindenberg, A., Straub, R. E., Lipska, B. K., et al. (2007). Genetic evidence implicating DARPP-32 in human frontostriatal structure, function, and cognition. *Journal of Clinical Investigation, 117*, 672–682.

Miller, E. K., & Cohen, J. D. (2001). An integrative theory of prefrontal cortex function. *Annual Review of Neuroscience, 24*, 167–202.

Mink, J. W. (1996). The basal ganglia: Focused selection and inhibition of competing motor programs. *Progress in Neurobiology, 50*, 381–425.

Moghaddam, B., & Adams, B. W. (1998). Reversal of phencyclidine effects by a group II metabotropic glutamate receptor agonist in rats. *Science, 281*, 1349–1352.

Murphy, K. C. (2002). Schizophrenia and velo-cardio-facial syndrome. *Lancet, 359*, 426–430.

Nicodemus, K., Kolachana, B., Vakkalanka, R., et al. (2007). Evidence for statistical epistasis between catechol-O-methyltransferase (COMT) and polymorphisms in RGS4, G72 (DAOA), GRM3, and DISC1: Influence on risk of schizophrenia. *Human Genetics, 120*, 889.

Nolan, K. A., Bilder, R. M., Lachman, H. M., & Volavka, J. (2004). Catechol-O-methyltransferase Val158Met polymorphism in schizophrenia: Differential effects of Val and Met alleles on cognitive stability and flexibility. *American Journal of Psychiatry, 161*, 359–61.

Norton, N., Williams, H., Dwyer, S., et al. (2005). No evidence for association between polymorphisms in GRM3 and schizophrenia. *BMC Psychiatry, 5*, 23.

O'Donovan, M. C., Craddock, N., Norton, N., et al. (2008). Identification of loci associated with schizophrenia by genome-wide association and follow-up. *Nature Genetics, 40*, 1053.

O'Reilly, R. C. (2006). Biologically based computational models of high-level cognition. *Science, 314*, 91–94.

O'Reilly, R. C., & Frank, M. J. (2006). Making working memory work: A computational model of learning in the prefrontal cortex and basal ganglia. *Neural Computation, 18*, 283–328.

Palmatier, M. A., Pakstis, A. J., Speed, W., et al. (2004). COMT haplotypes suggest P2 promoter region relevance for schizophrenia. *Molecular Psychiatry, 9*, 859.

Pantelis, C., Barnes, T. R., Nelson, H. E., et al. (1997). Frontal-striatal cognitive deficits in patients with chronic schizophrenia. *Brain, 120*, 1823–1843.

Patil, S. T., Zhang, L., Martenyi, F., et al. (2007). Activation of mGlu2/3 receptors as a new approach to treat schizophrenia: A randomized Phase 2 clinical trial. *Nature Medicine, 13*, 1102.

Petrides, M., & Pandya, D. N. (1984). Projections to the frontal cortex from the posterior parietal region in the rhesus monkey. *The Journal of Comparative Neurology, 228*, 105–116.

Pezawas, L., Meyer-Lindenberg, A., Goldman, A. L., et al. (2008). Evidence of biologic epistasis between BDNF and SLC6A4 and implications for depression. *Molecular Psychiatry, 13*, 709.

Ramsey, N. F., Koning, H. A., Welles, P., Cahn, W., Van Der Linden, J. A., & Kahn, R. S. (2002). Excessive recruitment of neural systems subserving logical reasoning in schizophrenia. *Brain, 125*, 1793–1807.

Robbins, T. W., & Roberts, A. C. (2007). Differential regulation of fronto-executive function by the monoamines and acetylcholine. *Cerebral Cortex, 17*, i151–160.

Rolls, E. T., Loh, M., Deco, G., & Winterer, G. (2008). Computational models of schizophrenia and dopamine modulation in the prefrontal cortex. *Nature Reviews Neuroscience, 9*, 696.

Sakai, K., & Passingham, R. E. (2002). Prefrontal interactions reflect future task operations. *Nature Neuroscience, 6*, 75–81.

Salthouse, T. A. (1996). The processing-speed theory of adult age differences in cognition. *Psychological Review, 103*, 403–428.

Sartorius, L. J., Nagappan, G., Lipska, B. K., et al. (2006). Alternative splicing of human metabotropic glutamate receptor 3. *Journal of Neurochemistry, 96*, 1139–1148.

Sawaguchi, T., & Goldman-Rakic, P. S. (1991). D1 dopamine receptors in prefrontal cortex: Involvement in working memory. *Science, 251*, 947–950.

Sawaguchi, T., & Goldman-Rakic, P. S. (1994). The role of D1-dopamine receptor in working memory: Local injections of dopamine antagonists into the prefrontal cortex of rhesus monkeys performing an oculomotor delayed-response task. *Journal of Neurophysiology, 71*, 515–528.

Seamans, J. K., & Yang, C. R. (2004). The principal features and mechanisms of dopamine modulation in the prefrontal cortex. *Progress in Neurobiology, 74*, 1–57.

Seamans, J. K., Durstewitz, D., Christie, B. R., Stevens, C. F., & Sejnowski, T. J. (2001a). Dopamine D1/D5 receptor modulation of excitatory synaptic inputs to layer V prefrontal cortex neurons. *PNAS, 98*, 301–306.

Seamans, J. K., Gorelova, N., Durstewitz, D., & Yang, C. R. (2001b). Bidirectional dopamine modulation of GABAergic inhibition in prefrontal cortical pyramidal neurons. *Journal of Neuroscience*, *21*, 3628.

Seasack, S. R., Hawrylak, V. A., Matus, C., Guido, M. A., & Levey, A. I. (1998). Dopamine axon varicosities in the prelimbic division of the rat prefrontal cortex exhibit sparse immunoreactivity for the dopamine transporter. *Journal of Neuroscience*, *18*, 2697–2708.

Selemon, L. D., & Goldman-Rakic, P. S. (1985). Longitudinal topography and interdigitation of corticostriatal projections in the rhesus monkey. *Journal of Neuroscience*, *5*, 776–794.

Shifman, S., Bronstein, M., Sternfeld, M., et al. (2002). A highly significant association between a COMT haplotype and schizophrenia. *The American Journal of Human Genetics*, *71*, 1296.

Silver, H., Feldman, P., Bilker, W. B., & Gur, R. C. (2003). Working memory deficit as a core neuropsychological dysfunction in schizophrenia. *American Journal of Psychiatry*, *160*, 1809–1816.

Tan, H. Y., Choo, W. C., Fones, C. S. L., & Chee, M. W. L. (2005). fMRI study of maintenance and manipulation processes within working memory in first-episode schizophrenia. *American Journal of Psychiatry*, *162*, 1849–1858.

Tan, H. Y., Sust, S., Buckholtz, J. W., et al. (2006). Dysfunctional prefrontal regional specialization and compensation in schizophrenia. *American Journal of Psychiatry*, *163*, 1969–1977.

Tan, H. Y., Callicott, J. H., & Weinberger, D. R. (2007a). Dysfunctional and compensatory prefrontal cortical systems, genes and the pathogenesis of schizophrenia. *Cerebral Cortex*, *17*, i171–i181.

Tan, H. Y., Chen, Q., Goldberg, T. E., et al. (2007b). Catechol-O-methyltransferase Val158Met modulation of prefrontal-parietal-striatal brian systems during arithmetic and temporal transformations in working memory. *Journal of Neuroscience*, *27*, 13393–13401.

Tan, H. Y., Chen, Q., Sust, S., et al. (2007c). Epistasis between catechol-O-methyltransferase and type II metabotropic glutamate receptor 3 genes in working memory brain function. *Proceedings of the National Academy of Sciences of the United States of America*, *104*, 12536–12541.

Tan, H. Y., Callicott, J. H., & Weinberger, D. R. (2008a). Intermediate phenotypes in schizophrenia genetics redux: Is it a no brainer? *Molecular Psychiatry*, *13*, 233–238.

Tan, H. Y., Nicodemus, K. K., Chen, Q., et al. (2008b). Genetic variation in AKT1 is linked to dopamine-associated prefrontal cortical structure and function in humans. *Journal of Clinical Investigation*, *118*, 2200–2208.

Tanaka, S. C., Doya, K., Okada, G., Ueda, K., Okamoto, Y., & Yamawaki, S. (2004). Prediction of immediate and future rewards differentially recruits cortico-basal ganglia loops. *Nature Neuroscience*, *7*, 887–893.

Toulopoulou, T., Picchioni, M., Rijsdijk, F., et al. (2007). Substantial genetic overlap between neurocognition and schizophrenia: Genetic modeling in twin samples. *Archives of General Psychiatry*, *64*, 1348–1355.

Vijayraghavan, S., Wang, M., Birnbaum, S. G., Williams, G. V., & Arnsten, A. F. T. (2007). Inverted-U dopamine D1 receptor actions on prefrontal neurons engaged in working memory. *Nature Neuroscience*, *10*, 376–84.

Wang, J., & O'Donnell, P. (2001). D1 dopamine receptors potentiate NMDA-mediated excitability increase in layer V prefrontal cortical pyramidal neurons. *Cerebral Cortex*, *11*, 452–462.

Wang, X.-J. (1999). Synaptic basis of cortical persistent activity: The importance of NMDA receptors to working memory. *Journal of Neuroscience*, *19*, 9587–9603.

Wei, Y., Williams, J. M., Dipace, C., et al. (2007). Dopamine transporter activity mediates amphetamine-induced inhibition of Akt through a $Ca2+$/calmodulin-dependent kinase II-dependent mechanism. *Molecular Pharmacology*, *71*, 835–842.

Weinberger, D. R., Egan, M. F., Bertolino, A., et al. (2001). Prefrontal neurons and the genetics of schizophrenia. *Biological Psychiatry*, *50*, 825–844.

Williams, G. V., & Goldman-Rakic, P. S. (1995). Modulation of memory fields by dopamine D1 receptors in prefrontal cortex. *Nature*, *376*, 572–575.
Williams-Gray, C. H., Hampshire, A., Robbins, T. W., Owen, A. M., & Barker, R. A. (2007). Catechol-O-methyltransferase val158met genotype influences frontoparietal activity during planning in patients with Parkinson's disease. *Journal of Neuroscience*, *27*, 4832–4838.
Winterer, G., & Weinberger, D. R. (2004). Genes, dopamine and cortical signal-to-noise ratio in schizophrenia. *Trends in Neurosciences*, *27*, 683–690.

COGNITIVE NEUROPSYCHIATRY
2009, 14 (4/5), 299–311

Psychology Press
Taylor & Francis Group

The role of general intelligence as an intermediate phenotype for neuropsychiatric disorders

Katherine E. Burdick[1,2,3], Nisali Gunawardane[2], Kristen Woodberry[2,4], and Anil K. Malhotra[1,2,3]

[1]*Feinstein Institute for Medical Research, Center for Translational Psychiatry, Manhasset, New York, USA,* [2]*Division of Psychiatry Research, Zucker Hillside Hospital, North Shore-Long Island Jewish Health System, Glen Oaks, New York, USA,* [3]*Albert Einstein College of Medicine, Bronx, New York, USA,* [4]*Department of Psychology, Harvard University, Cambridge, MA, USA*

Introduction. Neurocognitive impairment is common to several neuropsychiatric disorders. The growing use of cognitive impairment as an intermediate phenotype, or "endophenotype", in psychiatry raises the issue of whether global measures of cognition, such as IQ, or assays of more specific cognitive domains, such as working memory, will best serve to enhance power in detecting susceptibility loci in molecular genetic studies.
Methods. This paper will review the research on general intelligence in schizophrenia and bipolar disorder and evaluate its strengths and weaknesses as a candidate intermediate phenotype.
Results. Although global measures of cognition may not be optimal as intermediate phenotypes in bipolar disorder, certain clinical traits that overlap between schizophrenia and bipolar disorder, such as psychosis, may be predictive of poor performance on global measures, regardless of DSM-IV categorisation.
Conclusions. Global measures of cognition represent good intermediate phenotypes in schizophrenia. Current research does not support the use of global measures of cognition as intermediate phenotypes for bipolar disorder. Assays of specific neurocognitive domains may have greater potential to detect genetic markers for bipolar disorder.

Keywords: Bipolar disorder; Endophenotype; Genetic; Psychosis; Schizophrenia.

INTRODUCTION

Neurocognitive dysfunction is a core feature of several psychiatric disorders, including schizophrenia (Keefe & Fenton, 2007) and bipolar illness

Correspondence should be addressed to Katherine E. Burdick, Zucker Hillside Hospital, North-Shore-Long Island Jewish Health System, 75-59 263rd Street, Glen Oaks, NY 11004, USA. E-mail: kburdick@lij.edu

http://www.psypress.com/cogneuropsychiatry DOI: 10.1080/13546800902805347

(Balanzá-Martínez et al., 2008); however, the severity and pattern of cognitive impairment in these two disorders differs significantly (Daban et al., 2006), as does the timing of the onset of cognitive deficits. Specifically, deficits in global measures of cognition, such as intelligence quotient (IQ) or general cognitive ability composite scores (*g*), are very common in patients with schizophrenia (Keefe & Fenton, 2007) and are measurable prior to the onset of illness (Reichenberg et al., 2002), suggesting a genetic contribution to this trait in schizophrenia. In contrast, global cognitive impairment, less typical in patients with bipolar disorder (Torres, Boudreau, & Yatham, 2007), is noted only after the onset of the disease and may be restricted to acutely symptomatic periods (Balanzá-Martinez et al., 2008), suggesting that environmental, or disease-related features, may have more of a direct impact on IQ in bipolar disorder than risk factors related to genetics per se.

Despite differences in the nature or degree of global cognitive impairment in schizophrenia and bipolar disorder, several lines of evidence suggest that more specific aspects of cognition (i.e., working memory, executive function) may serve as useful "endophenotypes", or intermediate phenotypes, in molecular genetic studies of both illnesses (Bora, Yucel, & Pantelis, 2009; Braff, Freedman, Schork, & Gottesman, 2007). Gottesman and Gould (2003) articulated the concept of utilising an "endophenotype" in psychiatry as a biological marker to obtain "simpler clues to genetic underpinnings than the disease syndrome itself". Neurocognition has received increasing attention as a potential intermediate phenotype, as it appears to be closely linked with genetic susceptibility for several major mental illnesses and fulfils the basic "endophenotype" criteria previously set forth. Specifically, the four criteria that make a disease-related trait a good candidate as an intermediate phenotype for a given illness include the following: (1) it is consistently present in probands with the disorder; (2) it is not state related, but is measurable very early in the course of the disease and/or during periods of remission; (3) it is heritable; and (4) it is found, albeit to a lesser extent, in healthy (unaffected) relatives of probands when compared with unrelated healthy individuals (Gottesman & Gould, 2003).

With regard to the heritability of intelligence, there is a robust body of evidence that suggests that intelligence is significantly influenced by genetic factors (e.g., Deary, Spinath, & Bates, 2006; Plomin, 1999). Indeed, the intelligence quotient (IQ) is the most extensively studied cognitive domain with regard to heritability (Plomin & Spinath, 2004). IQ is measured with a battery of standardised tests, typically 2–4 for a quick estimate and 11–13 for a more robust measure. Standard IQ batteries (e.g., WAIS; Wechsler, 1981) include measures of verbal and nonverbal abilities, speeded and nonspeeded processing, and working memory. The number and composition of tests in an IQ estimate have relevance to genetic research, as there is some variability in the genetic versus environmental (e.g., education) contributions to the abilities measured by different tests (Deary et al., 2006). Of the four factors identified by confirmatory factor analysis of the WAIS-III subtests (verbal comprehension, perceptual organisation, working memory, and processing speed; Wechsler, 1997), verbal comprehension may be most highly heritable (heritability estimate of .84 relative to .63 to .68 for the other indices; Posthuma et al., 2003).

Premorbid IQ is often estimated with a single test, typically a test of word reading. Vocabulary and word recognition are not only highly correlated with "full scale" measures of IQ, they reflect what has been termed "crystallised" intelligence, or knowledge gleaned from experience. Especially in the case of single word reading, these abilities are relatively immune to the effects of illness, insult, or ageing, as long as these events have not themselves diminished years of education or other relevant experience. In contrast, tests such as the Raven's Progressive Matrices, also highly correlated with IQ and thus used to estimate IQ, have been considered measures of "fluid" intelligence, or the ability to reason and problem-solve abstractly and independent of prior learning or experience. Although more vulnerable to insult than word recognition, this more "fluid" intelligence is also highly heritable. In fact, in one study, roughly half of the variance in Raven's Progressive Matrices scores came from a genetic factor shared with all of the WAIS subtests (Rijsdijk, Vernon, & Boomsma, 2002).

This brings us to the psychometric construct of general cognitive ability (Spearman's *g*), first described in the early twentieth century (Spearman, 1904) and defined as the first factor of an unrotated principal component analysis (PCA). General cognitive ability *g* accounts for approximately 40% of the variance in performance on diverse neurocognitive measures (Jensen, 1998). The heritability of *g* has been established in multiple twin and family studies, with estimates generally ranging from 40% to 80% (Butcher, Davis, Craig, & Plomin, 2008). Although *g* differs somewhat from the notion of intelligence, as it is psychometrically defined and varies according to the specific battery by which it is determined, it is highly correlated with most measures of IQ and will be discussed here under the same rubric. However, to the degree that functioning within specific cognitive domains, such as memory or executive functioning, are genetically relevant to a specific disorder, composite scores or other *g* estimates from batteries in which these domains are included may have more endophenotypic promise than IQ.

In fact, recent attempts to parse out the genetics of more basic components of neurocognition have revealed that many specific neurocognitive domains have high degrees of heritability as well. Verbal ability, spatial ability, and memory also show substantial genetic influence, but less so than general intelligence (Plomin & DeFries 1998). More specifically, a strong genetic influence has been established in measures of attention (Fan, Wu, Fossella, & Posner, 2001), working memory (Dougherty et al., 2003), declarative memory (Finkel, Pedersen, McGue, & McClearn, 1995), and executive function (Swan & Carmelli, 2002), suggesting that these cognitive domains may also represent targets for use as intermediate phenotypes in molecular genetic studies. However, given the expectation that probands will demonstrate impairment, even during periods of remission, it is inherent that qualitative differences at the level of the cognitive phenotype in patient samples will be reflected by differences at the genetic level, such that genetic factors may play more or less of a role in different aspects of cognition in each disease. As such, certain cognitive domains may serve as ideal intermediate phenotypes in one psychiatric disorder, but not another, based on differing patterns of impairment in patients and their relatives. With this in mind, this manuscript will review the data supporting the use of cognition as an intermediate phenotype in schizophrenia and in bipolar disorder, with a specific focus on global measures of cognition (i.e., IQ, *g*).

SCHIZOPHRENIA

Cognitive impairments have long been recognised as a core characteristic of schizophrenia (Bleuler, 1950; Elvevåg & Goldberg, 2000), with significant, diffuse impairment that includes intellectual deterioration, as well as more specific deficits in individual cognitive domains, such as working memory and executive functioning (Keefe & Fenton, 2007). Thus, measures of both general and discrete cognitive functions may be useful as intermediate phenotypes in schizophrenia. We will focus here on the use of broad, global measures such as IQ.

Global cognitive deficits in schizophrenia probands

Deficits in intellectual functioning have been consistently reported in patients with schizophrenia. In a meta-analysis by Heinrichs and Zakzanis (1998) a large mean effect size for schizophrenia-healthy control differences was reported for measures of IQ (Cohen's $d = 1.10$). On average, the IQ scores of individuals with schizophrenia fell more than one standard deviation below those of healthy controls, with only 41.1% overlap in the two groups' distributions of scores.

Although the Heinrichs and Zakzanis (1998) meta-analysis was conducted on studies of chronic schizophrenia, recent data suggest a strikingly similar picture in schizophrenia patients experiencing their first-episode of psychosis, with large effect size differences noted in comparisons with healthy controls (Cohen's $d = 1.01$; Mesholam-Gately, Giuliano, Goff, Faraone, & Seidman, in press). These data suggest that intellectual deficits are present early in the course of the illness and cannot be explained solely by chronic symptoms or medication effects. In fact, IQ deficits are present during the premorbid period, as early as IQ is reliably measured, and long before the onset of psychotic symptoms (Cohen's $d = 0.54$; Woodberry, Giuliano, & Seidman, 2008). IQ impairments are therefore not only reliably associated with schizophrenia, but present very early in the developmental process and independent of clinical state.

Global cognitive deficits in relatives of schizophrenia probands

With the heightened interest in finding cognitive phenotypes that are more closely linked to the genetic susceptibility for schizophrenia, there have been numerous studies of neuropsychological functioning, including IQ, in the close relatives of patients with schizophrenia. The underlying premise of these family-based studies is that traits that are related to the expressed disorder will be present in the patient sample and absent in healthy relatives; in contrast, markers of genetic risk are expected to be present in both patients and close relatives when compared to unrelated healthy controls. In the majority of these studies relatives have demonstrated significant cognitive impairments in a number of specific domains, often strikingly parallel to those found in schizophrenia patient samples (e.g., Cannon et al., 1994; Kremen, Seidman, Pepple, Lyons, & Faraone, 1994). Given the heritability of both schizophrenia and IQ, it is perhaps not surprising that family studies have also found evidence of significantly greater IQ impairments in unaffected relatives of schizophrenia probands than in healthy controls (Cannon et al., 2000; Hughes et al., 2005; Kremen, Faraone, Seidman, Pepple, & Tsuang, 1998; McIntosh, Harrison, Forrester, Lawrie, & Johnstone, 2005).

It is important to note, however, that there have been a number of well-designed studies that have *not* found significant IQ impairment in schizophrenia relatives compared with controls (Goldberg et al., 1990, 1995; Toulopoulou, Quraishi, McDonald, & Murray, 2006). Methodological issues may help to explain discrepancies. For instance, Toulopoulou and colleagues (2006) tested a sample of relatives from families in which multiple members had schizophrenia and were thus at heightened genetic risk. Although they found no significant impairments in the relatives of schizophrenia patients compared with controls, the schizophrenia patients were selected to match a bipolar patient sample and appear to have been much less impaired than a typical schizophrenia sample. Indeed, a similar study of relatives at enhanced genetic risk, derived from multiplex families, did find significant intellectual impairment in this group (McIntosh et al., 2005).

Although results remain equivocal regarding the significance of IQ impairments in relatives of schizophrenia patients, the majority of studies reported IQ scores of relatives that were intermediate between healthy controls and schizophrenia probands. In addition, recent data derived from large-scale twin studies suggested that intellectual function was highly heritable within schizophrenia families (Greenwood et al., 2007; Toulopoulou et al., 2007). Further, using genetic model fitting to estimate the genetic relationship between cognition and schizophrenia, Toulopoulou et al. (2007) reported strong correlations between intelligence and schizophrenia, with shared genetic variance explaining more than 90% of the covariance between these traits. These data suggest that many of the same genes that contribute to variation in intelligence are also likely to influence susceptibility to schizophrenia, and vice versa (Toulopoulou et al., 2007). Indeed, several studies have now reported an association between markers within alleged schizophrenia susceptibility genes (i.e., dysbindin, *DTNBP1*) and general cognitive ability (*g*) (Burdick et al., 2006) and intelligence (Zinkstok et al., 2007).

BIPOLAR DISORDER

Historically, bipolar disorder has been differentiated from schizophrenia, at least in part, on the basis of a relatively complete interepisode recovery seen more often in the former than latter condition, particularly with regard to cognitive symptoms (Kraepelin, 1913). However, more recent data on neurocognition in bipolar disorder suggest that cognitive impairment is common and can be as severe as that noted in patients with schizophrenia, especially during acute episodes (Daban et al., 2006). Although the considerable effects of mood state on neurocognitive performance may confound data indicating neurocognitive impairment in acutely ill patients, patients with bipolar disorder also demonstrate neurocognitive impairment during periods of symptom remission. Specific deficits have been demonstrated in euthymic bipolar patients in the cognitive domains of attention, verbal learning, and executive function (for review, see Balanzá-Martinez et al., 2008; Torres et al., 2007).

Global cognitive deficits in bipolar probands

Global and diffuse neurocognitive impairment, consistent with deficits in IQ measures, has been reported during acute episodes of mania and depression, with more pronounced deficits on measures of Performance IQ than on Full-Scale IQ or Verbal IQ (Bearden, Hoffman, & Cannon, 2001; Daban et al., 2006). However, the majority of studies to date which have focused on identifying trait-like or stable cognitive dysfunction, do not report deficits in general cognitive ability, or IQ, in bipolar patients during periods of euthymia (Frangou, Haldane, Roddy, & Kumari, 2005; Pirkola et al., 2006; Robinson et al., 2006; Torres et al., 2007).

There are a few studies that do report significant differences in IQ measures in bipolar probands as compared with healthy controls; however, bipolar patients' performance is still typically noted to be within normal range according to normative data (Frantom, Allen, & Cross, 2008; McIntosh et al., 2005; Toulopoulou et al., 2006). Other moderating factors may also impact IQ in bipolar patients resulting in discrepancies across studies, including genetic susceptibility to psychosis. In a recent family study, McIntosh et al. (2005) evaluated IQ in a sample that included bipolar probands with a family history of bipolar disorder but also included bipolar probands with a family history of schizophrenia. When compared with healthy subjects with no family history of affective or psychotic illnesses, differences on IQ measures were only significant in bipolar probands derived from families with at least one first or second degree relative with bipolar disorder *and* one first or second degree relative with schizophrenia; whereas bipolar subjects with a family history of bipolar disorder but not schizophrenia performed comparable to controls (McIntosh et al., 2005). Similarly, data from a family study by Toulopoulou and colleagues (2006) reported impaired IQ in bipolar patients who were drawn from a sample that included only families with a history of psychosis.

As discussed, there are a handful of studies that have reported impairment in current IQ performance in bipolar patients; however, when *premorbid* intellectual capacity has been evaluated, bipolar patients have consistently demonstrated performance comparable to control subjects (Gourovitch et al., 1999; Kremen et al., 1998; Malhi et al., 2007; McIntosh et al., 2005; Toulopoulou et al., 2006). Consistent with this, a recent meta-analysis including 1446 euthymic bipolar patients concluded that premorbid intellectual capacity in bipolar patients did not differ from healthy controls, and although current IQ tended to be lower in the patient sample, significant heterogeneity between studies was noted (Bora et al., 2009). These data suggest that IQ deficits in bipolar patients are likely to reflect a decline in functioning due to the onset of the disease, and more specifically due to the onset of psychosis (Toulopoulou et al., 2006), suggesting a less substantial overlap between genetic susceptibility for bipolar disorder and IQ than that noted in patients with schizophrenia. Indeed, a genetic modelling study in families of bipolar probands found no significant relationship between genetic liability and measures of IQ (McIntosh et al., 2005).

Global cognitive deficits in relatives of bipolar probands

Current data suggest that global intelligence in relatives of bipolar patients is comparable to controls (Balanzá-Martinez et al., 2008). Although there remains a paucity of data from family studies in bipolar disorder, there have been several small-scale investigations of the impact that genetic risk may have on several aspects of cognition, including IQ. In a small study of seven monozygotic twin pairs discordant for bipolar disorder, both affected and unaffected cotwins demonstrated normal premorbid and current IQ as compared to the healthy control twins; however, significant differences were noted for measures of verbal learning, with unaffected twins performing significantly worse than control twins in this domain (Gourovitch et al., 1999). In a small sample of 15 unaffected offspring of bipolar I probands, no deficits were found in Full-Scale IQ as compared to a control sample; however specific aspects of executive functioning were significantly impaired in the high-risk relative group (Frangou et al., 2005).

As with the mixed results noted in euthymic bipolar patients, there have been some studies that reported significant differences in current IQ in first-degree relatives of bipolar probands versus healthy control samples, although IQ scores were well within the normal range based on normative data (Frantom et al., 2008; McIntosh et al., 2005; Toulopoulou et al., 2006). As previously discussed, a genetic liability to psychosis in these studies was predictive of lower IQ in both the probands and their relatives, suggesting a specific effect of psychosis and family history of psychosis on intellectual function. Finally, at least one study has reported significantly *higher* IQ in a sample of first-degree relatives of bipolar probands with a negative history of psychosis as compared with healthy controls (Kremen et al., 1998), suggesting that higher IQ in individuals with a genetic liability for bipolar disorder may serve as a protective factor against the development of the illness.

Overall, data from IQ studies in first-degree relatives of bipolar patients are equivocal; although intact premorbid functioning is a consistent finding. Reported impairments in current IQ in bipolar relatives may be confounded by sampling issues, including the use of above average healthy control samples. Intellectual differences between bipolar relatives and healthy controls appear to be modest, with performance typically falling within the normal range. Furthermore, the pattern of IQ deficits reported also appears be impacted by factors including a family history of schizophrenia or psychosis in the proband.

Thus, general intelligence and comparable global measures of neurocognitive ability probably do not represent good intermediate phenotypes in bipolar disorder as a group. Although outside of the scope of this review, other more specific aspects of neurocognition, including attention, verbal memory, and executive functioning appear to meet all of the basic criteria for use as an intermediate phenotype in bipolar disorder (Bora et al., 2009). Consistent with this, variation within at least one putative susceptibility loci for bipolar disorder, catechol-o-methyl-transferase (COMT) has been shown to influence verbal memory performance in bipolar patients, with no impact on IQ (Burdick et al., 2007).

CONCLUSION

Information derived from neurocognitive testing in neuropsychiatric disorders is undoubtedly useful for clinical planning, providing proxy measures of brain function by assessing both general cognitive ability and more discrete aspects of cognition such as attention, memory, and executive functions. The inclusion of cognitive measures in molecular genetic studies has recently gained popularity, at least in part due to the ease with which these tasks can be administered to the large samples necessary for genetic analyses. However, not all cognitive measures or domains are equivalent with regard to their appropriateness as intermediate phenotypes, or endophenotypes in the context of DSM-IV categorisation (American Psychiatric Association, 1994). Consideration for their use in this way should be made in a disease-specific manner.

In this review we suggest that deficits in general intelligence, and other global measures of cognition such as *g*, represent good potential candidates for intermediate phenotypes in studies of schizophrenia, as they are consistently noted in probands and are apparent long before the onset of frank psychosis (Reichenberg et al., 2002). Further, cognitive deficits are stable over the course of the illness (Keefe & Fenton, 2007) and are present in healthy relatives of patients with schizophrenia when compared with individuals who do not have increased genetic liability for the disease (Cannon et al., 1994). In contrast, the cognitive profiles of patients with bipolar illness and their unaffected relatives do not support the use of IQ as an intermediate phenotype for bipolar disorder, an illness for which more specific measures of neurocognitive functioning (i.e., verbal memory) or composites more reflective of these specific domains of functioning may have greater potential as genetic markers of the disease. Thus, nuances in the measurement of cognitive ability (e.g., number and type of domains included in IQ estimates or composite scores) are not immaterial to molecular genetic research. The degree to which a given measure is more or less sensitive to genetic versus environmental (e.g., education) effects has direct implications for the measure's usefulness as an endophenotypic marker.

However, as one considers the utility of neurocognition as an intermediate phenotype in molecular genetic studies, it may indeed be useful to do away with categorising patients by diagnosis and to consider subjects along more of a continuum of symptoms that overlap among several neuropsychiatric disorders. For example, psychosis is not unique to schizophrenia and it appears to be a key predictor of the extent of neurocognitive impairment, such that the presence of psychosis results in poorer cognitive function and a more generalised pattern of impairment that crosses diagnostic boundaries. Patients with schizophrenia, by definition, have psychotic features as part of their illness, but this feature is present in only approximately 50% of bipolar I patients. Notably, as compared with bipolar patients who never experience psychosis, bipolar patients with a history of psychosis are more likely to have IQ deficits and present with a pattern of neurocognitive dysfunction similar to that typically seen in schizophrenia. As genes do not code for diagnostic categories per se, but rather influence risk for a disorder via their effects on specific illness-related behaviours or symptoms, it is likely that the shared symptom of global neurocognitive impairment in patients with schizophrenia and with psychotic bipolar disorder reflects an overlap at the level of genetic liability to psychosis. Thus, it would be predicted that genes that have an influence on susceptibility to psychosis would be good candidates as genes for IQ, or general cognitive impairment, regardless of diagnostic categorisation. Future studies focused on finding genes that influence key traits associated with susceptibility to major mental illnesses might be better served by relying less on DSM-IV labels and more on the behavioural presentation of the patient group.

REFERENCES

American Psychiatric Association. (1994). *Diagnostic and statistical manual of mental disorders* (4th ed.). Washington, DC: Author.

Balanzá-Martinez, V., Rubio, C., Selva-Vera, G., Martinez-Aran, A., Sánchez-Moreno, J., Salazar-Fraile, J., et al. (2008). Neurocognitive endophenotypes (endophenocognitypes) from studies of relatives of bipolar disorder subjects: A systematic review. *Neuroscience Biobehavioral Reviews, 32*, 1426–1438.

Bearden, C. E., Hoffman, K. M., & Cannon, T. D. (2001). The neuropsychology and neuroanatomy of bipolar affective disorder: A critical review. *Bipolar Disorders, 3*, 106–150.

Bleuler, E. (1950). *Dementia praecox or the group of schizophrenias.* New York: International Universities Press.

Bora, E., Yucel, M., & Pantelis, C. (2009). Cognitive endophenotypes of bipolar disorder: A meta-analysis of neuropsychological deficits in euthymic patients and their first-degree relatives. *Journal of Affective Disorders, 113*, 1–20.

Braff, D. L., Freedman, R., Schork, N. J., & Gottesman, I. I. (2002). Deconstructing schizophrenia: An overview of the use of endophenotypes in order to understand a complex disorder. *Schizophrenia Bulletin, 33*, 21–32.

Burdick, K. E., Funke, B., Goldberg, J. F., Bates, J. A., Jaeger, J., Kucherlapati, R., & Malhotra, A. K. (2007). COMT genotype increases risk for bipolar I disorder and influences neurocognitive performance. *Bipolar Disorders, 9*, 370–376.

Burdick, K. E., Lencz, T., Funke, B., Finn, C. T., Szeszko, P. R., Kane, J. M., et al. (2006). Genetic variation in DTNBP1 influences general cognitive ability. *Human Molecular Genetics, 5*, 1563–1568.

Butcher, L. M., Davis, O. S. P., Craig, I. W., & Plomin, R. (2008). Genome-wide quantitative trait locus association scan of general cognitive ability using pooled DNA and 500K single nucleotide polymorphism microarrays. *Genes, Brain and Behavior, 7*, 435–446.

Cannon, T. D., Huttunen, M. O., Lonnqvist, J., Tuulio-Henriksson, A., Pirkola, T., Finkelstein, J., et al. (2000). The inheritance of neuropsychological dysfunction in twins discordant for schizophrenia. *American Journal of Human Genetics, 67*, 369–382.

Cannon, T. D., Zorrilla, L. E., Shtasel, D., Gur, R. E., Gur, R. C., Marco, E. J., et al. (1994). Neuropsychological functioning in siblings discordant for schizophrenia and healthy volunteers. *Archives of General Psychiatry, 51*, 651–661.

Daban, C., Martinez-Aran, A., Torrent, C., Tabarés-Seisdedos, R., Balanzá-Martínez, V., Salazar-Fraile, J., et al. (2006). Specificity of cognitive deficits in bipolar disorder versus schizophrenia. *Psychotherapy and Psychosomatics, 75*, 72–84.

Deary, I. J., Spinath, F. M., & Bates, T. C. (2006). Genetics of intelligence. *European Journal of Human Genetics, 14*, 690–700.

Dougherty, D. M., Bjork, J. M., Moeller, F. G., Harper, R. A., Marsh, D. M., Mathias, C. W., et al. (2003). Familial transmission of continuous performance test behavior: Attentional and impulsive response characteristics. *Journal of General Psychology, 130*, 5–21.

Elvevåg, B., & Goldberg, T. E. (2000). Cognitive impairment in schizophrenia is the core of the disorder. *Critical Reviews in Neurobiology, 14*, 1–21.

Fan, J., Wu, Y., Fossella, J. A., & Posner, M. I. (2001). Assessing the heritability of attentional networks. *BMC Neuroscience, 2*, 14–20.

Finkel, D., Pedersen, N. L., McGue, M., & McClearn, G. E. (1995). Heritability of cognitive abilities in adult twins: Comparison of Minnesota and Swedish data. *Behavior Genetics, 25*, 421–431.

Frangou, S., Haldane, M., Roddy, D., & Kumari, V. (2005). Evidence for deficit in tasks of ventral, but not dorsal, prefrontal executive function as an endophenotypic marker for bipolar disorder. *Biological Psychiatry, 58*, 838–839.

Frantom, L. V., Allen, D. N., & Cross, C. L. (2008). Neurocognitive endophenotypes for bipolar disorder. *Bipolar Disorders, 10*, 387–399.

Goldberg, T. E., Ragland, J. D., Torrey, E. F., Gold, J. M., Bigelow, L. B., & Weinberger, D. R. (1990). Neuropsychological assessment of monozygotic twins discordant for schizophrenia. *Archives of General Psychiatry, 47*, 1066–1072.

Goldberg, T. E., Torrey, E. F., Gold, J. M., Bigelow, L. B., Ragland, R. D., Taylor, E., et al. (1995). Genetic risk of neuropsychological impairment in schizophrenia: A study of monozygotic twins discordant and concordant for the disorder. *Schizophrenia Research, 17*, 77–84.

Gottesman, I. I., & Gould, T. D. (2003). The endophenotype concept in psychiatry: Etymology and strategic intentions. *American Journal of Psychiatry, 160*, 636–645.

Gourovitch, M. L., Torrey, E. F., Gold, J. M., Randolph, C., Weinberger, D. R., & Goldberg, T. E. (1999). Neuropsychological performance of monozygotic twins discordant for bipolar disorder. *Biological Psychiatry, 45*, 639–646.

Greenwood, T. A., Braff, D. L., Light, G. A., Cadenhead, K. S., Calkins, M. E., Dobie, D. J., et al. (2007). Initial heritability analyses of endophenotypic measures for schizophrenia: The consortium on the genetics of schizophrenia. *Archives of General Psychiatry, 64*, 1242–1250.

Heinrichs, W. R., & Zakzanis, K. K. (1998). Neurocognitive deficit in schizophrenia: A quantitative review of the evidence. *Neuropsychology, 12*, 426–445.

Hughes, C., Kumari, V., Das, M., Zachariah, E., Ettinger, U., Sumich, A., et al. (2005). Cognitive functioning in siblings discordant for schizophrenia. *Acta Psychiatrica Scandinavica, 111*, 185–192.

Jensen, A. R. (1998). *The g factor: The science of mental ability.* Westport, CT: Greenwood Publishing.

Keefe, R. S. E., & Fenton, W. S. (2007). How should DSM-V criteria for schizophrenia include cognitive impairment? *Schizophrenia Bulletin*, *33*, 912–920.
Kraepelin, E. (1913). *Psychiatrie: Vol. 8. Auflage*. Leipzig, Germany: J. A. Barth.
Kremen, W. S., Faraone, S. V., Seidman, L. J., Pepple, J. R., & Tsuang, M. T. (1998). Neuropsychological risk indicators for schizophrenia: A preliminary study of female relatives of schizophrenic and bipolar probands. *Psychiatry Research*, *79*, 227–240.
Kremen, W. S., Seidman, L. J., Pepple, J. R., Lyons, M. T., & Faraone, S. V. (1994). Neuropsychological risk indicators for schizophrenia: A review of family studies. *Schizophrenia Bulletin*, *20*(1), 103–119.
Malhi, G. S., Ivanovski, B., Hadzi-Pavlovic, D., Mitchell, P. B., Vieta, E., & Sachdev, P. (2007). Neuropsychological deficits and functional impairment in bipolar depression, hypomania, and euthymia. *Bipolar Disorders*, *9*, 114–125.
McIntosh, A. T., Harrison, L. K., Forrester, K., Lawrie, S. M., & Johnstone, E. C. (2005). Neuropsychological impairments in people with schizophrenia or bipolar disorder and their unaffected relatives. *British Journal of Psychiatry*, *186*, 378–385.
Mesholam-Gately, R., Giuliano, A. J., Goff, K. P., Faraone, S. V., & Seidman, L. J. (in press). Neurocognition in first episode schizophrenia: A meta-analytic review. *Neurocognition*.
Pirkola, T., Tuulio-Henriksson, A., Glahn, D., Kieseppa, T., Haukka, J., Kaprio, J., et al. (2005). Spatial working memory function in twins with schizophrenia and bipolar disorder. *Biological Psychiatry*, *58*, 930–936.
Plomin, R. (1999). Genetics and general cognitive ability. *Nature*, *402*, C25–C29.
Plomin, R., & DeFries, J. C. (1998). The genetics of cognitive abilities and disabilities. *Scientific American*, *278*, 62–69.
Plomin, R., & Spinath, F. M. (2004). Intelligence: Genetics, genes, and genomics. *Journal of Personality and Social Psychology*, *86*, 112–129.
Posthuma, D., Baaré, W. F. C., Hulshoff Pol, H. E., Kahn, R. S., Boomsma, D. I., & de Geus, E. J. C. (2003). Genetic correlations between brain volumes and the WAIS-III dimensions of verbal comprehension, working memory, perceptual organization, and processing speed. *Twin Research and Human Genetics*, *6*, 131–139.
Reichenberg, A., Weiser, M., Rabinowitz, J., Caspi, A., Schmeidler, J., Mark, M., et al. (2002). A population-based cohort study of premorbid intellectual, language, and behavioral functioning in patients with schizophrenia, schizoaffective disorder, and nonpsychotic bipolar disorder. *American Journal of Psychiatry*, *159*, 2027–2035.
Rijsdijk, F. V., Vernon, P. A., & Boomsma, D. I. (2002). Application of hierarchical genetic models to raven and WAIS subtests: A Dutch twin study. *Behavior Genetics*, *32*, 199–210.
Robinson, L. J., Thompson, J. M., Gallagher, P., Goswami, U., Young, A. H., Ferrier, I. N., & Moore, P. B. (2006). A meta-analysis of cognitive deficits in euthymic patients with bipolar disorder. *Journal of Affective Disorders*, *93*, 105–115.
Spearman, C. (1904). The proof and measurement of association between two things. *American Journal of Psychology*, *15*, 72–101.
Swan, G. E., & Carmelli, D. (2002). Evidence for genetic mediation of executive control: A study of aging male twins. *Journals of Gerontology: Psychological Sciences and Social Sciences*, *57B*, 133–143.
Torres, I. J., Boudreau, V. G., & Yatham, L. N. (2007). Neuropsychological functioning in euthymic bipolar disorder: A meta-analysis. *Acta Psychiatrica Scandinavica*, *434*(Suppl.), 17–26.
Toulopoulou, T., Picchioni, M., Rijsdijk, F., Hua-Hall, M., Ettinger, U., Sham, P., et al. (2007). Substantial genetic overlap between neurocognition and schizophrenia: Genetic modeling in twin samples. *Archives of General Psychiatry*, *64*, 1348–1355.
Toulopoulou, T., Quraishi, S., McDonald, C., & Murray, R. M. (2006). The Maudsley family study: Premorbid and current general intellectual function levels in familial bipolar I disorder and schizophrenia. *Journal of Clinical and Experimental Neuropsychology*, *28*, 243–259.

Wechsler, D. (1981). *Wechsler Adult Intelligence Scale-Revised*. San Antonio, TX: The Psychological Corporation.

Wechsler, D. (1997). *Wechsler Adult Intelligence Scale-Third Edition*. San Antonio, TX: The Psychological Corporation.

Woodberry, K. A., Giuliano, A. J., & Seidman, L. J. (2008). Premorbid IQ in schizophrenia: A meta-analytic review. *American Journal of Psychiatry, 165*, 579–587.

Zinkstok, J. R., de Wilde, O., van Amelsvoort, T. A., Tanck, M. W., Baas, F., & Linszen, D. H. (2007). Association between the DTNBP1 gene and intelligence: A case-control study in young patients with schizophrenia and related disorders and unaffected siblings. *Behavioral and Brain Functions, 20*, 3–19.

COGNITIVE NEUROPSYCHIATRY
2009, 14 (4/5), 312–331

Psychology Press
Taylor & Francis Group

Learning abilities and disabilities: Generalist genes in early adolescence

Oliver S. P. Davis[1], Claire M. A. Haworth[1], and Robert Plomin[1]

[1]*Social, Genetic & Developmental Psychiatry Centre, Institute of Psychiatry, King's College London, UK*

Introduction. The new view of cognitive neuropsychology that considers not just case studies of rare severe disorders but also common disorders, as well as normal variation and quantitative traits, is more amenable to recent advances in molecular genetics, such as genome-wide association studies, and advances in quantitative genetics, such as multivariate genetic analysis. A surprising finding emerging from multivariate quantitative genetic studies across diverse learning abilities is that most genetic influences are shared: they are "generalist", rather than "specialist".
Methods. We exploited widespread access to inexpensive and fast Internet connections in the United Kingdom to assess over 5000 pairs of 12-year-old twins from the Twins Early Development Study (TEDS) on four distinct batteries: reading, mathematics, general cognitive ability (g) and, for the first time, language.
Results. Genetic correlations remain high among all of the measured abilities, with language as highly correlated genetically with g as reading and mathematics.
Conclusions. Despite developmental upheaval, generalist genes remain important into early adolescence, suggesting optimal strategies for molecular genetic studies seeking to identify the genes of small effect that influence learning abilities and disabilities.

Keywords: Adolescence; Development; Genetics; Intelligence; Language; Learning ability; Mathematics; Reading; Twins.

Correspondence should be addressed to Oliver S. P. Davis, Social, Genetic & Developmental Psychiatry Centre, Institute of Psychiatry, King's College London, De Crespigny Park, London SE5 8AF, UK. E-mail: Oliver.Davis@iop.kcl.ac.uk

We thank the parents of the twins in TEDS and the twins themselves for making this study possible. TEDS is supported by a programme grant from the UK Medical Research Council (G500079), and this research is also supported by grants from the US National Institute of Child Health and Human Development (HD44454, HD46167, HD49861).

http://www.psypress.com/cogneuropsychiatry DOI: 10.1080/13546800902797106

INTRODUCTION

In the past, cognitive neuropsychology has tended to focus on case studies or relatively small samples (Caramazza & Coltheart, 2006), which limited the field's ability to take advantage of developments in genetics. Where genetic factors were considered, they were generally characterised as chromosomal or single-gene abnormalities, or innate species-wide processes. A new, broader view of cognitive neuropsychology is that it represents the cognitive level of analysis that lies between the brain and behaviour and aims to provide a full description and explanation, not just of normative species-wide processes and dramatic disruptions of these normal processes, but also of normal variation within species. Although chromosomal abnormalities and single-gene disorders in humans and genetic engineering studies in nonhuman species can be used to investigate genetics at the normative level of analysis, recent developments in genetics are most useful for investigating normal variation in human behaviour seen as quantitative traits and as common disorders that represent the extremes of these normal distributions (Plomin & Davis, 2009). For example, in molecular genetics, an advance that has revolutionised research is genome-wide association studies in which hundreds of thousands of common variations in DNA sequence are used to screen the genome for the DNA causes of genetic influence (Wellcome Trust Case Control Consortium, 2007). One consistent finding emerging from these studies is that complex traits and disorders are likely to be influenced by very many genetic variants, each of small effect size (McCarthy et al., 2008).

Quantitative genetic methods, such as inbred strain and selection studies in nonhuman animals and twin and adoption studies in the human species, emerged from the synthesis between Mendelian and biometric genetics more than a century ago (Fisher, 1918). After the rediscovery of Mendel's laws of inheritance in the early 1900s, Mendelians looked for single-gene effects seen in Mendelian segregation ratios, whereas biometricians argued that Mendel's laws could not apply to complex traits in plants or animals because these traits are distributed quantitatively, not qualitatively, and showed no simple pattern of inheritance. The resolution to this often bitter decade-long debate came with the realisation that Mendel's laws of inheritance of single genes also apply to complex traits if the traits are influenced by several genes, each of which are inherited according to Mendel's laws. With just a few genes, phenotypes begin to approach a normal distribution in the population. This notion that multiple-gene effects lead to quantitative traits is the cornerstone of quantitative genetic theory and methods (Falconer & MacKay, 1996; Fisher, 1918; Wright, 1921).

Quantitative genetic methods have primarily been used to discover the ubiquitous influence of genetics on normal variation at all levels of analysis from the brain to cognition to behaviour (Plomin, DeFries,

McClearn, & McGuffin, 2008). Importantly, quantitative genetics provides the "bottom line" of genetic influence on a trait regardless of how many genetic variants affect the trait or how small and complex their effects might be. Much remains to be learned even about this rudimentary question of *how much* genetics affects many cognitive and behavioural domains. However, the greatest impact of quantitative genetics will come from research that goes beyond this basic question to investigate *how* genes have their effect. A major example is multivariate genetic analysis, which investigates not only the variance of traits considered one at a time but also the covariance among traits. In this way, it indicates the extent to which the same or different genes affect several traits, using a statistic known as a *genetic correlation* (Neale, Boker, Xie, & Maes, 2006), which can be thought of as the probability that a gene associated with one trait is also associated with another trait. The genetic correlation can constrain explanations of cognitive neuropsychology. For example, it is reasonable to suppose that genetic effects will be specific to the substantially different cognitive processes involved in reading and mathematics, which would produce a low genetic correlation between the cognitive processes. A low genetic correlation indicating genetic specificity would lead to attempts to identify the genetically driven differences in brain processes that underlie these cognitive differences.

However, a very different result is emerging from multivariate genetic research on learning abilities and disabilities: Most genetic effects appear to be general in that the same genes affect different learning abilities and disabilities. A review of multivariate genetic research on learning abilities found that genetic correlations varied from .47 to .98 between reading and mathematics (three studies), from .67 to 1.0 between reading and language (five studies), and from .59 to .98 between language and mathematics (two studies) (Plomin & Kovas, 2005). The average genetic correlation was about .70, suggesting that a gene associated with reading, for example, would have a 70% chance of also being associated with mathematics or with language.

Moreover, the general effects of genes appear to extend beyond specific learning abilities such as reading and mathematics to other cognitive abilities such as verbal abilities (e.g., vocabulary and word fluency) and nonverbal abilities (e.g., spatial and memory). The average genetic correlation between specific learning abilities and general cognitive ability (*g*), which encompasses these verbal and nonverbal cognitive abilities, is about .60 (Plomin & Kovas, 2005). These findings have led to a *Generalist Genes* hypothesis (Plomin & Kovas, 2005), which has far-reaching implications for cognitive neuropsychology (Kovas & Plomin, 2006).

Although the Generalist Genes hypothesis has consistent support from multivariate genetic analyses, it was only recently tested by direct test measures in a sample large enough to conclusively establish the magnitude of the genetic correlations between learning abilities. In order to be able to test large numbers of individuals efficiently and inexpensively, we developed an online test battery that includes measures of reading, mathematics and *g*. One major advantage of this method of administration is that adaptive branching within each test allows the use of hundreds of items to test the full range of ability, while requiring individual children to complete only a relatively small number of items to ascertain their level of performance. We used this test battery to assess a UK-representative population sample of 2541 pairs of 10-year-old twins from the Twins Early Development Study (TEDS), by far the largest twin sample with cognitive test data (Haworth et al., 2007). Multivariate genetic analysis showed substantial genetic correlations between learning abilities: .57 between reading and mathematics, .61 between reading and *g*, and .75 between mathematics and *g*, providing strong support for the Generalist Genes hypothesis (Davis et al., 2008).

The purpose of the present study was to test the Generalist Genes hypothesis at age 12 in an even larger sample of twins. The difference between our previous analysis at age 10 and our present analysis at age 12 is greater than suggested by the interval of 2 years because 12 years marks the transition to adolescence, which involves brain changes (Ernst & Mueller, 2008; Spear, 2000) in addition to obvious hormonal changes. Underlying these physiological transformations are altered patterns of gene expression. Although changes in gene expression profiles do not necessarily entail changes in which DNA variants are associated with learning abilities, the upheaval may herald a shift in the relative importance of each of the variants, with some becoming more important while others, previously influential, become relatively ineffectual. These shifts have the potential to fundamentally affect the genetic architecture of learning abilities and disabilities at this age.

Moreover, three other improvements increased the scope of the present study to test the Generalist Genes hypothesis. First, at age 10, we assessed reading with a single test, whereas the present study at age 12 included a battery of four reading measures. Second, our previous study at age 10 did not include measures of language, nor has any other genetic research after infancy and early childhood. For this reason, we developed a language battery suitable for 12-year-olds that assesses both receptive spoken language and metalinguistic ability through three tests: syntactic, semantic, and pragmatic language; the language battery presents material orally to avoid confounding with reading ability, which is assessed through tests of reading comprehension, fluency, and accuracy (Haworth et al., 2007). A final improvement is the use of a latent factor approach in our model-fitting analyses. In our previous study, we created composite measures of mathematics and *g* and conducted an analysis of just three measures—reading, mathematics, and *g*. In contrast, in the present study we used a latent factor approach that included information from 14 tests, not just composite measures: four tests of reading, three tests of mathematics, three tests of language, and four tests of *g*. This latent factor approach made it possible to conduct more powerful multivariate genetic analyses at the level of the latent factors representing reading, mathematics, language, and *g* because the latent factors are independent of test-specific and uncorrelated error variance associated with each method of measurement.

Our hypothesis was that the Generalist Genes hypothesis would be supported despite the major transformations that occur during the transition from childhood to adolescence. Because we used a latent factor approach that excludes uncorrelated measurement error, we expected genetic correlations to be even greater in our present study at age 12 than in our previous study at age 10. Because language has not previously been included in tests of the Generalist Genes hypothesis after the language-learning era of early childhood, we had no hypothesis about the extent to which language conforms to the Generalist Genes hypothesis.

METHODS

Participants

TEDS recruited families of twins born in England and Wales in 1994, 1995, and 1996 (Oliver & Plomin, 2007). Since then, the sample has remained representative of the UK population (ascertained by comparison with census data from the Office of National Statistics; Kovas, Haworth, Dale, & Plomin, 2007). Although twins have the option of participating or not during each phase of data collection, the pairs that do participate remain representative of the larger sample. Informed consent is obtained by post or online consent forms, and a test administrator is then assigned who telephones the family and generally assists and encourages. Ethical approval for TEDS has been provided by the Institute of Psychiatry ethics committee, reference number 05/Q0706/228.

We excluded from the analyses children with severe current medical problems and children who had suffered severe problems at birth or whose mothers had suffered severe problems during pregnancy. We also excluded twins whose zygosity was unknown or uncertain or whose first language was other than English. Finally, we included only twins whose parents reported their ethnicity as "white", which is 93% of this UK sample. The present analyses are based on 5434 twin pairs (1945 monozygotic pairs, 1760 same-sex dizygotic, and 1729 opposite-sex dizygotic).

Measures

At age 12, the twins participated in Web-based testing. Widespread access to inexpensive and fast Internet connections in the UK has made online testing an attractive possibility for collecting data on the substantial samples necessary for genetic research, especially for multivariate genetic research. The advantages and potential pitfalls of data collection over the Internet have been reviewed in detail elsewhere (Birnbaum, 2004). For older children, most of whom are competent computer users, it is an interactive and enjoyable medium. As described earlier, adaptive branching allows the use of hundreds of items to test the full range of ability, while requiring individual children to complete only a relatively small number of items. In tests where it is appropriate, streaming voiceovers can minimise the necessary reading. In addition, the tests can be completed over a period of several weeks, allowing children to pace the activities themselves, although they are not allowed to return to items previously administered. Finally, it is possible to intersperse the activities with games. All of these factors help to maintain children's engagement with the tests. More details about the measures and their psychometric properties are available in Haworth et al. (2007).

General cognitive ability (g). The twins were tested on two verbal tests, WISC-III-PI Multiple Choice Information (General Knowledge) and Vocabulary Multiple Choice subtests (Wechsler, 1992), and two nonverbal reasoning tests, the WISC-III-UK Picture Completion (Wechsler, 1992) and Raven's Standard and Advanced Progressive Matrices (Raven, Court, & Raven, 1996; Raven, Court, & Raven, 1998), all administered online.

Reading. Four measures of reading ability were used: two measures of reading comprehension and a measure of reading fluency presented on the Web, and a fourth measure administered over the telephone.

Reading comprehension. The twins completed an adaptation of the reading comprehension subtest of the Peabody Individual Achievement Test (Markwardt, 1997), which we will refer to as $PIAT_{rc}$. The $PIAT_{rc}$ assesses literal comprehension of sentences. The sentences were presented individually on the computer screen. Children were required to read each sentence and were then shown four pictures. They had to select the picture that best matched the sentence they had read, using the mouse. All children started with the same items, but an adaptive algorithm modified item order and test discontinuation depending on the performance of the participant. The Internet adaptation of the $PIAT_{rc}$ was based on the same instructions, practice items, and test items as the original test.

As well as the $PIAT_{rc}$, we assessed reading comprehension using the GOAL Formative Assessment in Literacy for Key Stage 3 (GOAL plc., 2002). The GOAL is a test of reading achievement that is linked to the literacy goals for children at Key Stage 3 of the National Curriculum. Questions are grouped into three categories: Assessing Knowledge and Understanding (e.g., identifying information, use of punctuation and syntax), Comprehension (e.g., grasping meaning, predicting consequences), and Evaluation and Analysis (e.g., comparing and discriminating between ideas). Within each category, questions about words, sentences, and short paragraphs are asked. Because we were primarily interested in comprehension skills, we used questions from the two relevant categories, Comprehension, and Evaluation and Analysis, with 20 items from each category. Correct answers were summed to give a total comprehension score.

Reading fluency. Reading fluency was assessed using an adaptation of the Woodcock-Johnson III Reading Fluency Test (Woodcock, McGrew, & Mather, 2001) and the Test of Word Reading Efficiency (TOWRE, Form B; Torgesen, Wagner, & Rashotte, 1999). The Woodcock-Johnson is a measure of reading speed and rate that requires the ability to read and comprehend simple sentences quickly, e.g., "A flower grows in the sky?—Yes/No". The online adaptation consists of 98 yes/no statements; children need to indicate yes or no for each statement as quickly as possible. There is a time limit of 3 minutes for this test. Correct answers were summed to give a total fluency score.

The TOWRE, a standardised measure of fluency and accuracy in word reading skills, includes two subtests, each printed on a single sheet: a list of 85 words, called Sight-word Efficiency (SWE), which assesses the ability to read aloud real words; and a list of 54 nonwords, called Phonemic Decoding Efficiency (PDE), which assesses the ability to read aloud pronounceable printed nonwords. The child is given 45 s to read as many words as possible. Twins were individually assessed by telephone using test stimuli that had been posted to families in a sealed package with separate instructions that the package should not be opened until the time of testing. The same tester, who was blind to zygosity, assessed both twins in a pair within the same test session.

Mathematics. In order to assess mathematics, we developed an Internet-based battery that included questions from three components of mathematics. The items were based on the National Foundation for Educational Research 5–14 Mathematics Series, which is linked closely to curriculum requirements in the UK and the English Numeracy Strategy (NferNelson Publishing Co. Ltd., 1999). The presentation of items was streamed, so that items from different categories were mixed, but the data recording and branching were done within each category. The items were drawn from the following three categories: Understanding Number, Non-numerical Processes, and Computation and Knowledge. Understanding Number requires an understanding of the numerical and algebraic process to be applied when solving problems (e.g., understanding that multiplication and division are inverse operations). For example, "Look at the number 6085. Change the order of the figures around to make the biggest number possible". Another example is, "Type the missing number in the box: $27 + 27 + 27 + 27 + 27 + 27 = 27 \times$ ___". Non-numerical Processes do not rely solely on memory but rather require understanding of nonnumerical mathematical processes and concepts, such as rotational or reflective symmetry and other spatial operations. The questions do not have any significant numerical content that pupils need to consider. Three examples follow: "Which is the longest drinking straw? Click on it". "One of these shapes has corners that are the same. Click on this shape". "Which card appears the same when turned upside down? Click on it". Computation and Knowledge assesses the ability to perform straightforward computations using well-rehearsed pencil-and-paper techniques and the ability to recall mathematical facts and terminology. These questions either are mechanistic or rely on memorisation of mathematical facts and terminology. The operation is stated or is relatively unambiguous. Three examples follow: "Type in the answer: 76–39". "All 4-sided shapes are called? Click on the answer (Squares, Rectangles, Parallelograms, Kites, Quadrilaterals)". "Type in the answer: $149 + 785 = ?$". The mathematics battery is described in detail elsewhere (Kovas, Petrill, & Plomin, 2007).

Language. In order to assess receptive spoken language, standardised tests were selected that would discriminate children with language disability as well as being sensitive to individual differences across the full range of ability. Furthermore, an aspect of language that becomes increasingly important in adolescence—and which shows interesting variability at this age—is metalinguistic ability, which is knowledge about language itself (Nippold, 1998). For this reason, the three measures selected for testing included one with low metalinguistic demands designed to assess syntax (Listening Grammar) and two with higher demands that assess semantics (Figurative Language) and pragmatics (Making Inferences).

Syntax. Syntax was assessed using the Listening Grammar subtest of the Test of Adolescent and Adult Language (TOAL-3; Hammill, Brown, Larsen, & Wiederholt, 1994). This test requires the child to select two sentences that have nearly the same meaning, out of three options. The sentences are presented orally only.

Semantics. Semantics were assessed using Level 2 of the Figurative Language subtest of Test of Language Competence (Wiig, Secord, & Sabers, 1989), which assesses the interpretation of idioms and metaphors; correct understanding of such nonliteral language requires rich semantic representations. The child hears a sentence orally and chooses one of four answers, presented in both written and oral form.

Pragmatics. Level 2 of the Making Inferences subtest of the Test of Language Competence (Wiig et al., 1989) assessed an aspect of pragmatic language, requiring participants to make permissible inferences on the basis of existing (but incomplete) causal relationships presented in short paragraphs. The child hears the paragraphs orally and chooses two of four responses, presented in both written and oral form.

Statistical analyses

According to the quantitative genetic model (Plomin et al., 2008), twins reared together resemble each other due to the additive effects of shared genes (A) or shared (common) environmental factors (C). For identical or monozygotic (MZ) twins, the correlation between their genes is 1.00, whereas for nonidentical or dizygotic (DZ) twins, the correlation is .50 because DZ twins on average share half of their segregating alleles. The correlation between twins for shared environment is, by definition, 1.00 for both MZ and DZ twins growing up in the same family, whereas nonshared environmental influences (E) are uncorrelated and contribute to differences between twins. For the twin analyses, standardised residuals correcting for age and sex were used because the age of twins is perfectly correlated across pairs, which means that, unless corrected, variation within each age group at the time of testing would contribute to the correlation between twins and be misrepresented as shared environmental influence. The same applies to the sex of the twins, since MZ twins are always of the same sex. The assumptions of the classical twin model, and their validity, have been discussed in detail elsewhere (Boomsma, Busjahn, & Peltonen, 2002; Visscher, Hill, & Wray, 2008).

As well as examining twin correlations in R (http://www.r-project.org), we used standard ACE model-fitting analysis in Mx (Neale et al., 2006) where ACE stands for additive genetic influences (*A*), shared or common environmental influences (*C*), and nonshared environmental (*E*) influences, as before. Model-fitting analysis specifies a correlational structure (a model) using matrix algebra. This model is a hypothesis about the structure of the dataset, and is derived from what we know about how MZ and DZ twins are related to each other (see earlier). By fitting the model to the data using an iteration process, we can assess its "goodness of fit" and estimate the contributions of *A*, *C*, and *E*.

To explore shared genetic and environmental aetiology, we fitted a *common pathway* model to raw data (Figure 1; Neale et al., 2006). This model derives latent factors for each domain using maximum-likelihood factor analysis. It fixes the variance of these latent factors at 1 and partitions them into *A*, *C*, and *E* components. It also partitions the covariance between the latent factors in the same way. Similarly, residual variance at each age is

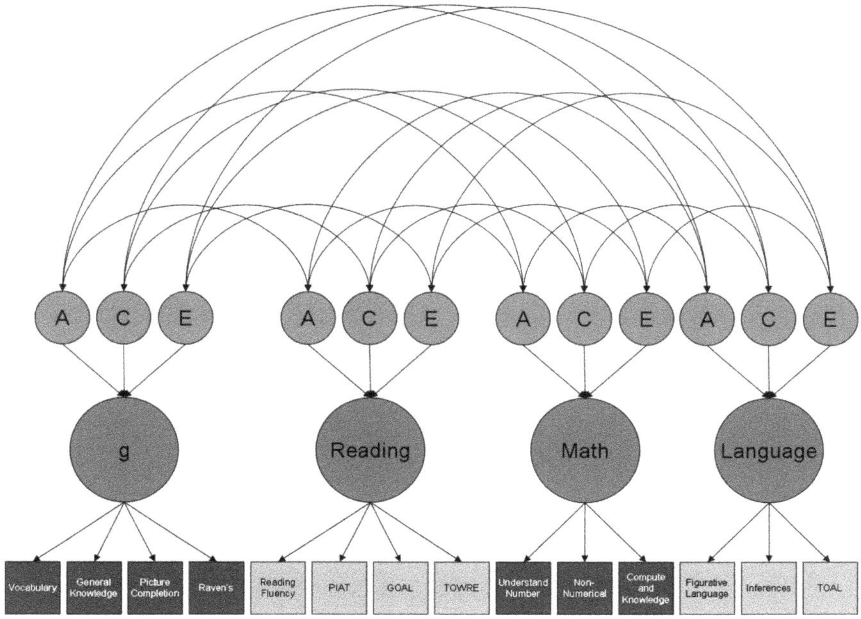

Figure 1. Common pathway model. A = additive genetic effects; C = shared (common) environmental effects; E = nonshared environmental effects. Squares represent measured traits; circles represent latent factors. The lower tier of arrows represents factor loadings; the second tier represents genetic and environmental path coefficients; the curved arrows at the top represent correlations between genetic and environmental latent factors.

partitioned into *A*, *C*, and *E* components (not shown). Earlier studies indicated very little difference in *ACE* estimates between males and females (Kovas, Haworth, et al., 2007), implying no significant differences in aetiology, so we combined DZ same-sex and DZ opposite-sex twin pairs for the individual differences analysis. Even though the genetic and environmental aetiology of the sexes is similar, this does not preclude differences in mean performance for males and females.

RESULTS

Phenotypic analyses

Table 1 presents the measure means and standard deviations, subdivided by sex and zygosity. It also presents the results of an analysis of variance testing the effects of sex and zygosity on the measures. Although our large sample gives us around 90% power to detect a difference in means as small as 0.1 standard deviations, in each measure age and sex accounted for less than 3%

TABLE 1
Measure means and standard deviations by sex and zygosity

		Male		*Female*		*MZ*		*DZ*		*ANOVA*				
Battery	*Measure*	*M*	*SD*	*M*	*SD*	*M*	*SD*	*M*	*SD*	*Sex*	*Zyg.*	*Sex ∗ Zyg.*	R^2	*N pairs*
g	Vocabulary	0.03	0.96	−0.01	1.01	−0.05	1.02	0.04	0.97	.18	.02	.46	.00	4001
	General knowledge	0.19	0.97	−0.15	1.01	−0.10	1.01	0.06	1.00	.00*	.00*	.05	.03	4285
	Picture completion	0.03	1.03	−0.02	0.97	−0.05	1.01	0.04	0.99	.04	.00	.15	.00	3846
	Raven's	−0.01	1.03	0.00	0.99	−0.03	1.00	0.01	1.02	.81	.28	.97	.00	4149
Reading	Reading fluency	−0.06	0.99	0.07	1.01	−0.05	1.00	0.05	1.00	.04	.00*	.35	.01	4757
	PIAT	0.06	1.00	−0.04	1.00	−0.07	1.01	0.05	0.99	.00	.00*	.22	.01	4875
	GOAL	0.02	1.03	0.01	0.96	−0.03	1.02	0.04	0.98	.10	.00*	.04	.00	4711
	TOWRE	−0.01	1.02	0.01	0.98	−0.04	1.00	0.03	0.99	.79	.02	.28	.00	4061
Mathematics	Understand number	0.09	1.00	−0.08	1.00	−0.04	0.99	0.02	1.01	.00*	.22	.80	.01	4627
	Non-numerical	0.04	1.01	−0.03	0.99	−0.03	1.00	0.02	1.01	.07	.19	.66	.00	4655
	Compute and knowledge	0.05	1.03	−0.03	0.98	−0.03	1.00	0.03	1.00	.02	.06	.26	.00	4624
Language	Figurative language	0.05	0.98	0.00	0.99	−0.01	0.98	0.04	0.99	.23	.19	.72	.00	4136
	Inferences	−0.02	1.02	0.04	0.99	0.00	1.01	0.02	1.00	.54	.28	.44	.00	3953
	TOAL	−0.06	0.99	0.03	1.00	−0.06	0.99	0.02	1.00	.22	.02	.50	.00	3849

M = mean; *SD* = standard deviation; MZ = monozygotic twins; DZ = dizygotic twins; Sex = *p*-value for sex effect; Zyg. = *p*-value for zygosity effect; Sex ∗ Zyg. = *p*-value for sex by zygosity interaction; R^2 = proportion of total variance accounted for by sex and zygosity; Analysis of variance performed using one member of each twin pair; *Significant at $p = .05$ after a Bonferroni correction for multiple testing ($p < .00357$).

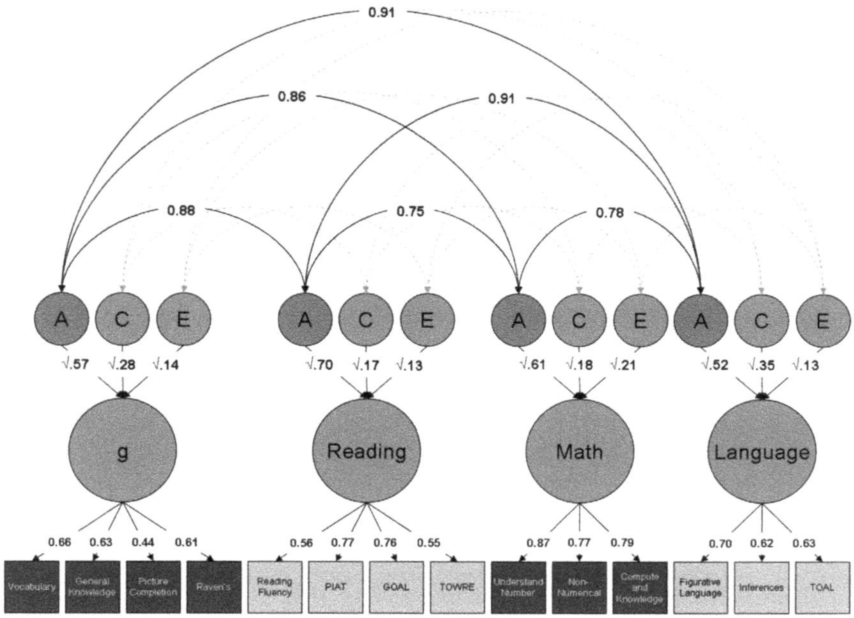

Figure 2. Cross-trait genetic effects. Estimates of cross-trait additive genetic effects (A) are highlighted.

of the variance, and more often less than 1% (indexed by the coefficient of determination, R^2). Factor loadings of the individual measures onto the latent factors (the bottom of Figure 2) are consistently high, ranging from .44 to .87 with a mean of .67. They are also generally similar, indicating that each constituent measure contributes similarly to the factor.

Univariate genetic analyses

Intraclass correlations (twin similarity coefficients) are presented in Table 2 for the MZ and DZ twins at each age. Correlations between MZ twins were consistently higher than those between DZ twins, suggesting a genetic contribution to each measure. The significance of genetic influence is indicated by the lack of overlap between the confidence intervals for MZ and DZ twins. As a first estimate of the effect size (heritability), doubling the difference between the MZ and DZ correlations yields heritability estimates ranging from 22% to 74% with a mean of 39%. Shared environmental influences are estimated as the extent to which MZ resemblance exceeds heritability: They range from 1% to 26%, with a mean of 14%. The

TABLE 2
Twin (intraclass) correlations

Battery	*Measures*	*MZ*	*N pairs*	*DZ*	*N pairs*
g	Vocabulary	.43 (.38–.47)	1421	.25 (.21–.29)	2389
	General knowledge	.56 (.52–.59)	1500	.41 (.38–.44)	2524
	Picture completion	.46 (.41–.50)	1367	.31 (.27–.34)	2275
	Raven's	.48 (.44–.52)	1456	.29 (.26–.33)	2474
Reading	Reading fluency	.72 (.70–.75)	1667	.42 (.38–.45)	2844
	PIAT	.55 (.52–.58)	1694	.35 (.31–.38)	2950
	GOAL	.49 (.45–.52)	1635	.32 (.29–.35)	2812
	TOWRE	.75 (.73–.77)	1488	.38 (.35–.41)	2540
Mathematics	Understand number	.58 (.55–.61)	1573	.40 (.36–.43)	2709
	Non-numerical	.55 (.51–.58)	1596	.35 (.32–.38)	2734
	Compute and knowledge	.54 (.51–.58)	1581	.30 (.27–.34)	2712
Language	Figurative language	.53 (.49–.57)	1462	.35 (.32–.39)	2455
	Inferences	.40 (.36–.44)	1398	.29 (.25–.33)	2351
	TOAL	.44 (.40–.48)	1377	.29 (.25–.33)	2312

95% confidence intervals in parentheses.

remainder of the variance is attributed to nonshared environmental influences (plus error of measurement): ranging from 28% to 60%, with a mean of 47%.

Common pathway model

Figures 2 and 3, and 4 depict results for the common pathway model used to investigate the shared aetiology. As explained previously, this model partitions variance into latent factors representing *g*, reading, mathematics, and language, and residual variance specific to each measure. The variance is then further partitioned into additive genetic (*A*), shared (common) environmental (*C*), and nonshared environmental (*E*) influences. Confidence intervals for the estimates in Figures 2 and 3, and 4 are presented in Table 3.

Heritability of the latent factors (a^2) is consistently high, accounting for more than half of the variance: 57% for *g*, 70% for reading, 61% for mathematics, and 52% for language. The remainder of the variance in the latent factors is accounted for by environmental influences, split between the shared (common) environment (c^2)—28%, 17%, 18%, and 35%—and the nonshared (unique) environment (e^2): 14%, 13%, 21%, and 13%.

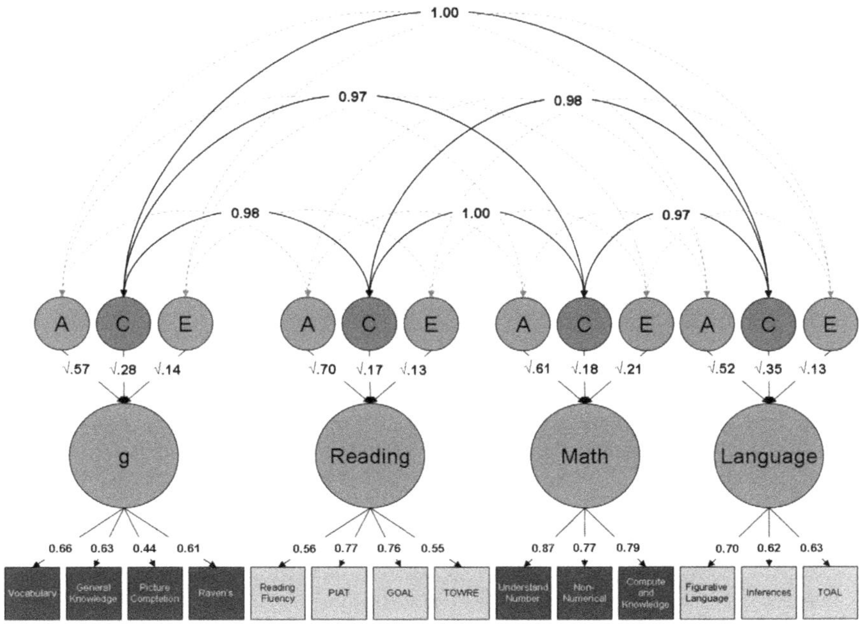

Figure 3. Cross-trait shared environmental effects. Estimates of cross-trait shared environmental effects (C) are highlighted.

Specific variance components (i.e., the variance not accounted for by the latent factors) are shown alongside the latent factor variance components in Table 3. The specific heritability is consistently low, ranging from nearly 0 to .03, with a mean of .013, indicating that all the genetic variance on these measures is subsumed in the latent factors. Specific shared environment is also low (.00 to .02, with a mean of .0064). In contrast, specific nonshared environment ranges from .21 to .54, with a mean of .39. As well as measure-specific nonshared environmental effects, this component partly represents measure-specific measurement error that is not subsumed into the latent factor for each domain.

Genetic correlations between latent factors are uniformly high, ranging from .75 to .91, with a mean of .85 (Figure 2). The shared environmental correlations are even higher, ranging from .97 to 1.00, with a mean of .98 (Figure 3). Finally, nonshared environmental correlations are moderate, ranging from .50 to .94, with a mean of .62. Interestingly, the nonshared environmental correlation between *g* and language is substantially higher than the other nonshared environmental correlations: .94, compared to a mean of .55 (Figure 4).

TABLE 3
Common pathway model estimates

Measures		*Common* A	*Common* C	*Common* E	
g		.57 (.49–.57)	.28 (.21–.28)	.14 (.12–.17)	
Reading		.70 (.63–.74)	.17 (.11–.19)	.13 (.11–.14)	
Mathematics		.61 (.53–.67)	.18 (.12–.20)	.21 (.19–.24)	
Language		.52 (.41–.55)	.35 (.25–.40)	.13 (.09–.17)	
		Specific A	*Specific* C	*Specific* E	*Factor loading*
g	Vocabulary	.02 (.01–.02)	.01 (.00–.01)	.51 (.48–.53)	.66 (.64–.67)
	General knowledge	.01 (.01–.02)	.00 (.00–.00)	.40 (.37–.43)	.63 (.62–.64)
	Picture completion	.00 (.00–.00)	.00 (.00–.00)	.52 (.48–.55)	.44 (.42–.46)
	Raven's	.01 (.01–.02)	.00 (.00–.00)	.44 (.41–.45)	.61 (.60–.63)
Reading	Reading fluency	.00 (.00–.00)	.00 (.00–.00)	.25 (.23–.27)	.56 (.54–.56)
	PIAT	.01 (.01–.01)	.01 (.00–.01)	.37 (.34–.39)	.77 (.75–.77)
	GOAL	.01 (.01–.01)	.01 (.01–.01)	.40 (.38–.42)	.76 (.75–.77)
	TOWRE	.00 (.00–.00)	.00 (.00–.00)	.21 (.19–.22)	.55 (.53–.56)
Mathematics	Understand number	.03 (.01–.03)	.02 (.02–.03)	.22 (.21–.24)	.87 (.86–.87)
	Non-numerical	.01 (.01–.02)	.02 (.01–.02)	.32 (.29–.33)	.77 (.75–.78)
	Compute and knowledge	.01 (.01–.02)	.02 (.02–.02)	.30 (.28–.32)	.79 (.78–.80)
Language	Figurative language	.03 (.03–.03)	.00 (.00–.00)	.40 (.37–.41)	.70 (.69–.72)
	Inferences	.02 (.01–.03)	.00 (.00–.00)	.54 (.50–.57)	.62 (.60–.64)
	TOAL	.02 (.02–.03)	.00 (.00–.00)	.51 (.47–.52)	.63 (.61–.64)

95% confidence intervals in parentheses; sample-size adjusted Bayesian Information Criterion = −180921.

The genetic and environmental correlations are summarised in Table 4, alongside the bivariate heritability and environmentality of the latent factors. Whereas the genetic correlation indexes the genetic overlap between the latent factors independent of the heritabilities of the latent factors, the bivariate heritability indexes the proportion of the phenotypic correlation between the latent factors that is mediated by genetic effects. Genetic effects consistently account for over half of the phenotypic correlation, ranging from 53% to 65%, with a mean of 61%. Shared environment accounts for just over a quarter of the phenotypic correlation, ranging from 23% to 34%, with a mean of 28%. Although the shared environmental correlation is even higher than the genetic correlation, the mediation of the phenotypic correlation is lower because the shared environment accounts for a smaller proportion of the variance of each of the latent factors. Finally, the nonshared environment accounts for the remainder of the phenotypic correlation, contributing from 8% to 14%, with a mean of 11%.

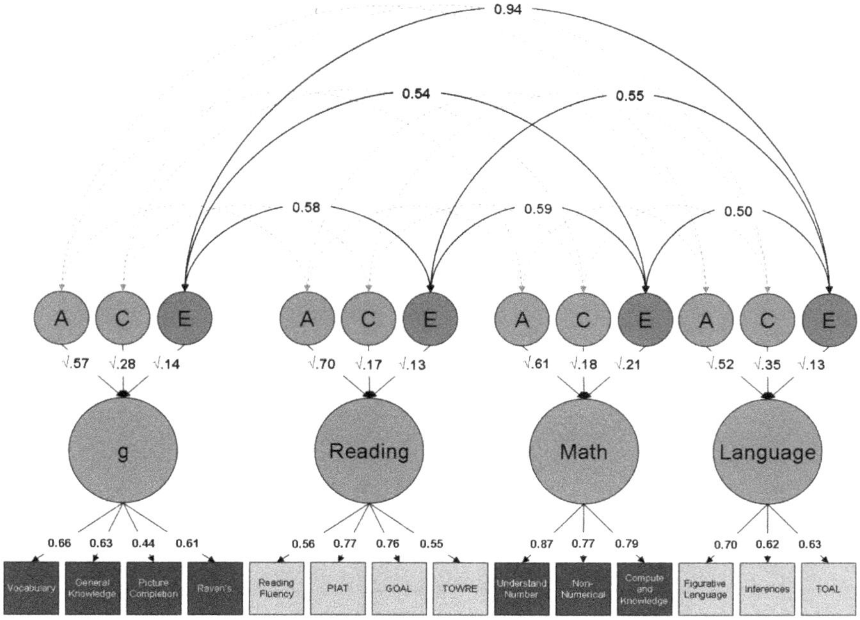

Figure 4. Cross-trait nonshared environmental effects. Estimates of cross-trait nonshared environmental effects (E) are highlighted.

DISCUSSION

Generalist genes

The high heritability of our latent factors confirms that genetic effects continue to be important in the aetiology of cognitive abilities and disabilities into early adolescence. However, going beyond this, the high genetic correlations between the latent factors representing *g*, reading, mathematics, and language confirm that the genetic influences on these domains continue to be largely shared, in spite of accompanying hormonal and brain changes, with genes accounting for most of the phenotypic correlation between these domains. This implies that when genes influencing reading, for example, are found, they are also very likely to be associated with mathematics, general cognitive ability and language. These correlations are consistent with various possible causal pathways (Kovas & Plomin, 2006); so far we do not have the evidence to distinguish between them, but future research combining genetics with neuropsychiatry will bring us closer to uncovering the mechanisms through which these genes have their generalist influence. Finding strong genetic correlations suggests that, in our search for genes influencing learning abilities, we should not be looking

TABLE 4
Cross-trait correlations and bivariate estimates from common pathway model

	Correlations			
Measures	r_A	r_C	r_E	r_P
g and reading	.88 (.84–.88)	.98 (.88–.99)	.58 (.48–.68)	.85 (.83–.86)
g and mathematics	.86 (.81–.90)	.97 (.87–1.00)	.54 (.45–.60)	.82 (.80–.83)
g and language	.91 (.87–.94)	1.00 (.95–1.00)	.94 (.80–1.00)	.94 (.93–.95)
Reading and mathematics	.75 (.71–.75)	1.00 (.90–1.00)	.59 (.52–.67)	.76 (.75–.78)
Reading and language	.91 (.86–.96)	.98 (.88–1.00)	.55 (.43–.55)	.86 (.84–.87)
Mathematics and language	.78 (.73–.78)	.97 (.87–1.00)	.50 (.41–.55)	.77 (.75–.78)
	Mediation of r_P			
	$a_x a_y r_A / r_P$	$c_x c_y r_C / r_P$	$e_x e_y r_E / r_P$	
g and reading	.65 (.58–.72)	.25 (.19–.31)	.09 (.07–.09)	
g and mathematics	.62 (.54–.67)	.27 (.20–.32)	.11 (.09–.14)	
g and language	.53 (.45–.60)	.34 (.26–.39)	.14 (.11–.16)	
Reading and mathematics	.64 (.56–.70)	.23 (.16–.29)	.13 (.11–.15)	
Reading and language	.64 (.57–.66)	.28 (.21–.34)	.08 (.06–.83)	
Mathematics and language	.57 (.50–.65)	.32 (.25–.37)	.11 (.09–.13)	

r_A = genetic correlation; r_C = shared environmental correlation; r_E = nonshared environmental correlation; r_P = phenotypic correlation; $a_x a_y r_A / r_P$ = proportion of phenotypic correlation accounted for by genetic factors; $c_x c_y r_C / r_P$ = proportion accounted for by shared environmental factors; $e_x e_y r_E / r_P$ = proportion accounted for by nonshared environmental factors; 95% confidence intervals in parentheses.

for genetic variants influencing a specific trait independent of other traits, because there will be relatively few such genes. Instead, we should be seeking the generalist genes and exploring how they interact developmentally with the environment through changes in gene expression and methylation to influence diverse brain and cognitive outcomes. Despite this, a mean genetic correlation of .85 is not the same as a correlation of 1.00; although the vast majority of genes are generalist, there are likely to be a few that remain specific to one domain or another. Finding these genes will be an even greater challenge than identifying the generalist genes, because they account for a smaller proportion of the genetic variance.

Although the current paper considers the entire distribution of variation, including the low end of the distribution in *g*, reading, mathematics, and language, our previous research has shown that genetic factors influencing variation in the normal range also influence the extremes (Kovas, Haworth, et al., 2007), so in addition to genes being generalist between domains, they also make little distinction between ability and disability; in terms of cognition, the abnormal is normal.

Shared environment

Although the shared environment accounted for a relatively small proportion of the variance in each of our domains, and mediated only around a quarter of the phenotypic correlation, the very high shared environmental correlations (higher even than the genetic correlations) imply that the shared environmental influences on each of the traits are almost entirely the same. Although there has been some success, identifying the specific shared environments that influence cognition is difficult, in part because what appears to be an environmental effect is often partly genetically mediated, through the process of gene–environment correlation (Jaffee & Price, 2007; Plomin & Davis, 2006). This correlation can be passive, where an individual inherits both genes and the childhood family environment from a parent; active, where an individual seeks out particular environments influenced by their genetic propensities; or evocative, where genetically influenced behaviour leads an individual to evoke certain reactions from others in the environment. For this reason, if we are to identify the specific environments that contribute to the development of cognition, it is best to couch the study within a genetically sensitive design. It is likely that when we identify the shared environments—those that make children growing up in the same family more similar—they will be substantially the same for *g*, reading, mathematics, and language.

Nonshared environment

The nonshared environment has relatively little influence on our latent factors because uncorrelated environments and measurement error are not included in the factor. This variance appears as measurement-specific variance instead, in contrast to the genetic and shared environmental influences where almost all of the variance is subsumed into the latent factor. Similarly, although the nonshared environment is still substantially correlated across traits, it is less correlated than genetic and shared environmental factors, implying that it is largely nonshared environments that bring about unevenness in cognitive ability profiles. However, there is one exception: The nonshared environmental correlation between *g* and language is much higher than the correlations among the other abilities. Because this is the first study to explore the aetiology of language in relation to other cognitive abilities at this age, this is an exciting finding; it will be interesting to see whether this pattern continues as the TEDS twins progress through adolescence. If so, it could have important implications for our understanding of the development of human cognition and how it relates to the development of language, both throughout childhood and throughout the course of evolution.

CONCLUSION

By testing *g*, reading, mathematics, and language in a large sample of 12-year-old twins, we have demonstrated that the Generalist Genes hypothesis continues to hold true into early adolescence. This has implications affecting the hunt for the many genes of small effect that are expected to influence human cognitive abilities, and also our conception of the pathway from genes to brain to behaviour. By understanding that the same genes act similarly on quite different domains, we can begin to explore the mechanisms by which this genetic variation influences the biochemical pathways and neurological networks involved in the day-to-day life of the brain. In addition, we have made an unexpected finding—the nonshared environment, which so often differentiates performance in different domains, appears to act similarly on general cognitive ability and language performance in early adolescence: a clear reminder that both genes and dynamic environments are important influences in the development of human cognition.

REFERENCES

Birnbaum, M. H. (2004). Human research and data collection via the Internet. *Annual Review of Psychology, 55*, 803–832.

Boomsma, D., Busjahn, A., & Peltonen, L. (2002). Classical twin studies and beyond. *Nature Reviews Genetics, 3*, 872–882.

Caramazza, A., & Coltheart, M. (2006). Cognitive neuropsychology twenty years on. *Cognitive Neuropsychology, 23*(1), 3–12.

Davis, O. S. P., Kovas, Y., Harlaar, N., Busfield, P., McMillan, A., Frances, J., et al. (2008). Generalist genes and the Internet generation: Etiology of learning abilities by web testing at age 10. *Genes, Brain and Behavior, 7*, 455–462.

Ernst, M., & Mueller, S. C. (2008). The adolescent brain: Insights from functional neuroimaging research. *Developmental Neurobiology, 68*(6), 729–743.

Falconer, D. S., & MacKay, T. F. C. (1996). *Introduction to quantitative genetics* (4th ed.). Harlow, UK: Longman.

Fisher, R. A. (1918). The correlation between relatives on the supposition of Mendelian inheritance. *Transactions of the Royal Society of Edinburgh, 52*, 399–433.

GOAL plc. (2002). *GOAL formative assessment: Key stage 3*. London: Hodder & Stoughton.

Hammill, D. D., Brown, V. L., Larsen, S. C., & Wiederholt, J. L. (1994). *Test of Adolescent and Adult Language (TOAL-3)*. Austin, TX: Pro-Ed.

Haworth, C. M. A., Harlaar, N., Kovas, Y., Davis, O. S. P., Oliver, B. R., Hayiou-Thomas, M. E., et al. (2007). Online cognitive testing of large samples needed in genetic research. *Twin Research and Human Genetics, 10*(4), 554–563.

Jaffee, S. R., & Price, T. S. (2007). Gene–environment correlations: A review of the evidence and implications for prevention of mental illness. *Molecular Psychiatry, 12*(5), 432–442.

Kovas, Y., Haworth, C. M. A., Dale, P. S., & Plomin, R. (2007). The genetic and environmental origins of learning abilities and disabilities in the early school years. *Monographs of the Society for Research in Child Development, 72*, 1–144.

Kovas, Y., Petrill, S. A., & Plomin, R. (2007). The origins of diverse domains of mathematics: Generalist genes but specialist environments. *Journal of Educational Psychology, 99*(1), 128–139.

Kovas, Y., & Plomin, R. (2006). Generalist genes: Implications for the cognitive sciences. *Trends in Cognitive Sciences, 10*, 198–203.

Markwardt, F. C. Jr. (1997). *Peabody Individual Achievement Test–Revised (normative update) manual.* Circle Pines, MN: American Guidance Service.

McCarthy, M. I., Abecasis, G. R., Cardon, L. R., Goldstein, D. B., Little, J., Ioannidis, J. P. A., & Hirschhorn, J. N. (2008). Genome-wide association studies for complex traits: Consensus, uncertainty and challenges. *Nature Reviews Genetics, 9*, 356–369.

Neale, M. C., Boker, S. M., Xie, G., & Maes, H. H. (2006). *Mx: Statistical modeling* (7th ed.). Department of Psychiatry, VCU Box 900126, Richmond, VA 23298, USA.

NferNelson Publishing Co. Ltd. (1999). *Mathematics 5–14 series.* Windsor, UK: Author.

Nippold, M. A. (1998). *Later language development: The school-age and adolescent years.* Austin, TX: Pro-Ed.

Oliver, B. R., & Plomin, R. (2007). Twins Early Development Study (TEDS): A multivariate, longitudinal genetic investigation of language, cognition and behavior problems from childhood through adolescence. *Twin Research and Human Genetics, 10*, 96–105.

Plomin, R., & Davis, O. S. P. (2006). Gene–environment interactions and correlations in the development of cognitive abilities and disabilities. In J. MacCabe, O. O'Daly, R. M. Murray, P. McGuffin, & P. Wright, (Eds.), *Beyond nature and nurture in psychiatry: Genes, environment and their interplay* (pp. 35–46). Oxford, UK: Informa Healthcare Medical Books.

Plomin, R., & Davis, O. S. P. (2009). The future of genetics in psychology and psychiatry: Microarrays, genome-wide association, and non-coding RNA. *Journal of Child Psychology and Psychiatry, 50*(1–2), 63–71.

Plomin, R., DeFries, J. C., McClearn, G. E., & McGuffin, P. (2008). *Behavioral genetics* (5th ed.). New York: Worth.

Plomin, R., & Kovas, Y. (2005). Generalist genes and learning disabilities. *Psychological Bulletin, 131*, 592–617.

Raven, J. C., Court, J. H., & Raven, J. (1996). *Manual for Raven's Progressive Matrices and Vocabulary Scales.* Oxford, UK: Oxford University Press.

Raven, J. C., Court, J. H., & Raven, J. (1998). *Manual for Raven's Advanced Progressive Matrices.* Oxford, UK: Oxford Psychologists Press.

Spear, L. P. (2000). The adolescent brain and age-related behavioral manifestations. *Neuroscience and Biobehavioral Reviews, 24*, 417–463.

Torgesen, J. K., Wagner, R. K., & Rashotte, C. A. (1999). *Test of Word Reading Efficiency.* Austin, TX: Pro-Ed.

Visscher, P. M., Hill, W. G., & Wray, N. R. (2008). Heritability in the genomics era—concepts and misconceptions. *Nature Reviews Genetics, 9*, 255–266.

Wechsler, D. (1992). *Wechsler Intelligence Scale for Children manual* (3rd ed., UK; WISC-III-UK). London: Psychological Corporation.

Wellcome Trust Case Control Consortium. (2007). Genome-wide association study of 14,000 cases of seven common diseases and 3,000 shared controls. *Nature, 447*, 661–678.

Wiig, E. H., Secord, W., & Sabers, D. (1989). *Test of Language Competence* (Expanded ed.). San Antonio, TX: Psychological Corporation.

Woodcock, R. W., McGrew, K. S., & Mather, N. (2001). *Woodcock-Johnson III Tests of Achievement.* Itasca, IL: Riverside Publishing.

Wright, S. (1921). Systems of mating. *Genetics, 6*, 111–178.

COGNITIVE NEUROPSYCHIATRY
2009, 14 (4/5), 332–355

Psychology Press
Taylor & Francis Group

Genetic variation influences on the early development of reactive emotions and their regulation by attention

Brad E. Sheese[1], Pascale Voelker[2], Michael I. Posner[3], and Mary K. Rothbart[3]

[1]*Department of Psychology, Illinois Wesleyan University, Bloomington, IN, USA,* [2]*Institute of Molecular Biology, University of Oregon, Oregon, USA,* [3]*Department of Psychology, University of Oregon, Oregon, USA*

Introduction. Individual differences in temperament and attention provide an important link between normal and pathological development. Previous studies suggest that during infancy, orienting of attention is associated with higher levels of positive affect and lower levels of negative affect. For older children and adults, self-regulation, as measured by ratings of effortful control, is consistently associated with lower levels of negative affect such as sadness and distress.
Methods. In the current paper, we use a longitudinal study of children at ages 6–7 months (Time 1) and 18–20 months (Time 2) to examine how variations in candidate genes relate to emotional and self-regulatory aspects of temperament.
Results. In accord with previous findings, parent ratings of orienting were positively related to positive affect only during infancy. Genetic variation in COMT was related to positive affect at Time 1 but not Time 2. Negative affect at both Time 1 and Time 2 was related to genetic variation in SNAP25. Genetic variation in CHRNA4 was related to Effortful Control at Time 2.
Conclusions. These findings lend support to the early modulation of emotion by aspects of orienting (Time 1) and executive attention (Time 2), and indicate that emotional reactivity and its regulation are modulated by different genes.

Keywords: Attention; Orienting; Genes; Emotional-reactivity; Self-regulation.

Correspondence should be addressed to Brad E. Sheese, Department of Psychology, Illinois Wesleyan University, Bloomington, IN 61701, USA. E-mail: bsheese@iwu.edu

This paper was prepared for a special issue of *Cognitive Neuropsychiatry* on Genes and Cognition. It reports some results for a longitudinal study starting with 7-month-old infants, continued at 18–20 months and ending fall 2008 at about 4 years of age. The study was supported by NIH Grant HD 38051 to Georgia State University and by the Brain Biology and Machine Initiative at the University of Oregon with funding from the Telemedicine and advanced technology research center (TATRC). Manuscript preparation was supported in part by a grant from Illinois Wesleyan University.

Reprints from bsheese@iwu.edu, or mposner@uoregon.edu

http://www.psypress.com/cogneuropsychiatry DOI: 10.1080/13546800902844064

INTRODUCTION

A temperamental approach to development and psychopathology is concerned with constitutionally based dispositions that underlie individual differences in patterns of development. Temperament describes differences in emotional reactivity as well as cognitive processes that contribute to self-regulation, such as attention. Differences in reactive emotions (fear, smiling and laughter) that arise early in infancy are described by two separate temperament dimensions. Differences in positive emotions are grouped under a higher level temperamental factor called surgency or extraversion, while negative affectivity is a factor summarizing fear, frustration and other negative emotions. As development proceeds these two reactive systems are increasingly regulated by networks related to attention. Early in development, individual differences in the efficiency of orienting attention underlie infant's abilities to regulate their own emotional states. As more advanced forms of attention, such as executive attention, improve, they become critical to the development of self-regulation and form a core aspect of individual differences in temperament called effortful control (Posner & Rothbart, 2000, 2007). Highlighting the importance of the development of this aspect of temperament, research has linked individual differences in effortful control to a number of forms of psychopathology, including conduct disorder and ADHD (Nigg, Friderici, Park, & Zucker, 2007; Rothbart, 2007; Rothbart & Posner, 2006).

Rothbart and associates have summarized events related to the development of emotional reactivity (positive and negative) and their regulation (Rothbart, 2007; Rothbart & Bates, 2006). In infancy there is limited self-regulation; instead regulation is provided by the caregiver, often by orienting the infant to various environmental events that can compete with distress or stimulate positive affect. Orienting thus serves as a major regulatory influence on emotion by enhancing positive emotion and suppressing negative affect. Consistent with this idea, during infancy positive affect has a substantial correlation with the duration of orienting of attention (Putnam, Rothbart, & Gartstein, 2008) and orienting has been shown to temporarily lower the expression of negative affect (Harman, Rothbart, & Posner, 1997). Emotions are regulated differently in infancy from their later regulation in older children and adults. For older children and adults, effortful control, related to self-regulation via executive attention mechanisms, operates to suppress negative affect. In the United States, effortful control, as measured by self or parent report, is most often negatively correlated with reports of negative affect (Rothbart & Rueda, 2005).

Historically theorists have attempted to link aspects of temperament to specific underlying properties of the nervous system (Rothbart & Bates, 2006). Behavioral genetics research using twins has generally supported substantial genetic contributions to temperament including both positive and negative affect (Saudino, 2005). Recent efforts have used the candidate gene approach to associate temperament with specific allelic variations in genes (Posner, Rothbart, & Sheese, 2007). This approach is promising because it links individual differences in temperament to specific variations in brain function or structure. However, the candidate gene approach can also be problematic, because effect sizes tend to be small and results are sometimes difficult to replicate. Psychological traits, such as temperament, are thought to be influenced by multiple genes and multiple environmental factors so the effect size associated with any given single genetic variation is likely to be small. For example, extensive research on trait intelligence has shown genetic effect sizes may be as small as .1% (Plomin, Kennedy, & Craig, 2006). One way to address this issue is to examine polymorphisms that have shown a pattern of associations with psychological phenotypes in previous research.

Dopamine plays an important role in the regulation of cognitive and affective behavioral processes. Individual differences in dopamine neurotransmission can bias normal and pathological variability in associated behaviors (Esposito, DiMatteo, & Di Giovanni, 2008; Forbes, Brown, Kimak, Ferrell, Manuck, & Hariri, 2007). The DRD4 exon III repeat polymorphism has been linked to novelty seeking in adults, intensity of reaction in toddlers (De Luca et al., 2003), and adaptability at 1–5 months of age (De Luca et al., 2001). The DRD4 polymorphism, in association with the 5HTTLPR polymorphism, was correlated with negative emotionality in 2-month-old infants (Auerbach et al., 1999) and responses to novelty in 12 month olds (Lakatos et al., 2003). DRD4 has been related to sensation seeking and activity level, aspects of positive affect or surgency (Auerbach, Faroy, Ebstein, Kahana, & Levine, 2001).

In an earlier paper from our longitudinal study (Sheese, Voelker, Rothbart, & Posner, 2007) we found that children without the 7 repeat allele of the DRD4 gene were little influenced by quality of parenting on behavioral dimensions related to sensation seeking, while those with the 7 repeat showed higher than normal sensation seeking only if they had lower quality parenting. Despite relatively small Ns this same result has been found in several investigations using a variety of parenting and outcome measures (Bakermans-Kranenburg & van IJzendoorn, 2006; van IJzendoorn & Bakermans-Kranenburg, 2006). One study used a parental intervention and random assignment to experimental and control conditions. The intervention was more effective in children with the 7 repeat than those without it, thus supporting a causative role for the interaction between gene and environment (Bakersmans-Kranenburg, Van IJzendoorn, Pijlman, Mesman, & Juffer, 2008). Although these studies provide evidence that there are consistent genetic effects that can be examined by small N studies, it is clearly important to obtain replication to confirm findings of any one study. The use of specific hypotheses can also help to compensate for multiple comparisons in these studies.

Research on DRD4 and other dopamine-related genes implicate the dopaminergic pathway in the regulation of surgency, positive affectivity, and novelty seeking. Research also shows that the dopaminergic pathway is crucial to many aspects of attention. Consequently, polymorphisms in genes that directly or indirectly regulate or mediate dopamine neurotransmission appear to be good candidates for studying temperament. In the current study we examined polymorphisms that directly influence dopaminergic signaling, such as COMT, MAOA, DAT1, and DRD4, and polymorphisms in genes that indirectly modulate dopamine neurotransmission, such as CHRNA4, TPH2, 5HTT, APOE, BDNF, and SNAP25, in relation to temperament in infancy and early childhood.

Inconsistencies in the current literature may be related to the ages studied. For example, during childhood the serotonin transporter gene (5-HTTLPR) has been most frequently linked to aspects of negative affect (Caspi et al., 2002). However, in adults it appears related to Extraversion or positive affect (Gillihan, Farah, Sankoorikal, Breland, & Brodkin, 2008). We expect that similar differences may arise because, unlike genotype, gene expression, phenotype, experience and self-regulation change over the life span. A longitudinal study provides one vehicle for attempting to disentangle some of these factors.

In order determine the relation of genetic variations to early development of reactive emotions and self-regulation we have been conducting a longitudinal study with children aged 6–7 months to 4 years (Sheese, Rothbart, Posner, White, & Fraundorf, 2008). To date we have tested the children as infants (Time 1, 6–7 months) and toddlers (Time 2, 18–20 months) and genotyped them for candidate genes related to early emotion or self-regulation. A full list of the genes and polymorphisms used appears in Table 1.

In the current paper we examine four specific hypotheses. The first hypothesis is that genes related to positive affect will differ from those related to negative affect. This difference could be most prominent during infancy where the observed emotion may depend more clearly on reactivity rather than on self-regulation. A second hypothesis is that during infancy, positive and negative affect will be influenced by orienting. The orienting network arises early in development and may serve as the primary regulatory system until the executive network is more fully developed (Posner & Rothbart, 2007). Third, when effortful control develops in childhood it will serve to moderate negative affect as has been found throughout later development (Rothbart & Rueda, 2005). Fourth, genes that relate to more reactive emotion will differ from those related to self-regulation based on the different brain networks involved in reactive emotion from those involved in self-regulation.

TABLE 1
Candidate genes and polymorphism-based comparison groups

Candidate genes	*Polymorphism*	*Polymorphism-based comparison groups*		
CHRNA4	rs1044396	C/C	T/C	T/T
COMT	rs4680	G/G	A/G	A/A
DBH	rs1108580	G/G	A/G	A/A
MAOA*	30 bp VNTR (1)	3/3-repeat	3/4-repeat	4/4-repeat (Sabol et al., 1998)
SNAP25 MnlI	rs3746544	G/G	G/T	T/T
Dichotomized candidate genes				
5HTT (SLC6A4)	44 bp ins/del (2)	short present	short absent (Deckert et al., 1997)	
APOE	rs429358 and rs7412	TT or CT present	TT or CT absent	
BDNF	rs6265	A (met) present	A absent	
DAT1(SLC6A3)	rs28363170 40 bp VNTR	9-repeat present	9-repeat absent	
DRD4, 7-repeat	48 bp VNTR (3)	7-repeat present	7-repeat absent (Van Toll et al., 1992)	
DRD4, 4-repeat	48 bp VNTR (3)	4-repeat present	4-repeat absent	
SNAP25 DdeI	rs1051312	C present	C absent	
TPH2	rs4570625	G present	G absent	

*5-repeat excluded.

METHOD

Data for the current study are from an ongoing longitudinal study of the development of attention and temperament conducted at the University of Oregon. Children were first brought into the lab when they were 6–7 months of age (Time 1 assessment). They returned again when the children were 18–20 months of age (Time 2 assessment). To offset participants lost to attrition, additional participants were recruited into the study at Time 2. The study will conclude with an assessment of attention and temperament related outcomes when the children are 4 years of age. Methods and data reported here are from both Time 1 and Time 2 assessments.

Participants

Fifty infants (20 females) were initially recruited through birth announcements or fliers distributed to day care centers and parenting programs in Eugene, Oregon and surrounding communities. These flyers informed participants of the longitudinal nature of the study and the focus on attention, emotion and behavior. During the original recruitment no genetic studies were contemplated and thus this aspect of the study was not important in recruitment. Participants were predominantly white and non-Hispanic. In the initial data collection with this sample, the majority of infants were either 6 months ($N = 21$) or 7 months ($N = 23$) of age; however, three infants were slightly younger than 6 months and three infants were slightly older than 7 months (age range, 5.86–8.16 months; M age in days = 209.86; $SD = 19.51$).

Of the original 50 Time 1 participants, 37 participants returned at Time 2 and completed the ECBQ. These 37 participants are used for longitudinal analyses of temperament stability. Additional families were recruited to supplement the Time 2 sample. A total of 70 families participated at Time 2. Fifty-four families agreed to provide samples for genotyping. Of these, eight did not return the temperament questionnaire. The final sample used for Time 2 analyses presented here included 46 children with complete ECBQ data and genotyping information (16 females and 29 males; 85% white and non-Hispanic, 10% Asian, 5% unreported or other).

Procedure

Families were mailed temperament questionnaires prior to arriving at the laboratory. At Time 1 and Time 2 families came into the lab for a session lasting less than 1 h. In these sessions parent and child participated in a variety of laboratory assessments designed to examine different aspects of cognitive and social development. During the Time 2 laboratory session buccal cell samples were collected for genotyping. The current paper focuses on the relation between child's genetics and parent-reported temperament. Previous publications (Sheese et al., 2007, 2008) describe measures and results from other parts of the study that are not included in the current paper.

Temperament – Time 1

Caregivers were asked to complete the Infant Behavior Questionnaire – Revised (IBQ-R; Gartstein & Rothbart, 2003). Descriptive statistics and alphas for the IBQ were computed using all available data. The IBQ-R contains 14 sub-scales that can be organized into three higher-order factors. The first higher-order factor is Positive Affect ($M = 4.77$, $SD = .61$, $\alpha = .76$). Positive affect subscales include: Approach (rapid approach, excitement, and positive anticipation of pleasurable activities; $M = 5.24$, $SD = .90$, $\alpha = .84$), Smiling and Laughter (smiling or laughter from the child in general caretaking and play situations; $M = 4.80$, $SD = .99$, $\alpha = .84$), Vocal Reactivity (vocalization exhibited by the baby in daily activities; $M = 4.61$, $SD = .92$, $\alpha = .83$), High Intensity Pleasure (pleasure or enjoyment related to high stimulus intensity, rate, complexity, novelty, and incongruity; $M = 5.86$, $SD = .63$, $\alpha = .80$), Activity Level (movement of arms and legs, squirming and locomotor activity; $M = 4.12$, $SD = .82$, $\alpha = .80$) and Perceptual Sensitivity (detection of slight, low intensity stimuli from the external environment; $M = 3.97$, $SD = 1.07$, $\alpha = .82$).

The second higher-order factor is Negative Affect ($M = 3.09$, $SD = .66$, $\alpha = .75$). Negative affect subscales include: Sadness (general low mood; lowered mood and activity specifically related to personal suffering, physical state, object loss, or inability to perform a desired action; $M = 3.48$, $SD = .86$, $\alpha = .87$), Distress to Limitations (fussing, crying or showing distress while (a) in a confining place or position; (b) involved in caretaking activities; or (c) unable to perform a desired action; $M = 3.56$, $SD = .77$, $\alpha = .77$), Fear (startle or distress to sudden changes in stimulation, novel physical objects or social stimuli; inhibited approach to novelty; $M = 2.42$, $SD = .98$, $\alpha = .91$), and Falling Reactivity/Rate of Recovery from Distress (reversed; rate of recovery from peak distress, excitement, or general arousal; ease of falling asleep; $M = 5.12$, $SD = .85$, $\alpha = .83$).

The third higher-order factor is Orienting ($M = 4.99$, $SD = .53$, $\alpha = .51$). Orienting subscales include: Duration of Orienting (attention to and/or interaction with a single object for extended periods of time; $M = 3.86$, $SD = 1.15$, $\alpha = .83$), Low Intensity Pleasure (amount of pleasure or enjoyment related to situations involving low stimulus intensity, rate, complexity, novelty, and incongruity; $M = 5.29$, $SD = .85$, $\alpha = .83$), Soothability (reduction of fussing, crying, or distress when the caretaker uses soothing techniques; $M = 5.07$, $SD = .55$, $\alpha = .64$), and Cuddliness (expression of enjoyment and molding of the body to being held by a caregiver; $M = 5.73$, $SD = .64$, $\alpha = .87$).

Temperament – Time 2

Caregivers were asked to complete the Early Childhood Behavior Questionnaire (ECBQ; Putnam, Gartstein & Rothbart, 2006). Descriptive statistics and alphas for the ECBQ were computed using all available data. The ECBQ consists of 18 subscales that can be organized into three higher order factors. The first ECBQ higher-order factor is Positive Affect ($M = 4.32$, $SD = .38$, $\alpha = .65$), consisting of the following subscales: Activity Level/Energy (rate and intensity of gross motor activity, including rate and extent of locomotion; $M = 4.94$, $SD = .72$, $\alpha = .74$), High-intensity Pleasure (pleasure or enjoyment related to situations involving high stimulus intensity, rate, complexity, novelty and incongruity; $M = 4.97$, $SD = .72$, $\alpha = .67$), Impulsivity (speed of response initiation; $M = 5.09$, $SD = .71$, $\alpha = .70$), Motor Activation (repetitive small-motor movements, fidgeting; $M = 2.15$, $SD = .68$, $\alpha = .52$), Positive Anticipation (excitement about expected pleasurable activities; $M = 4.63$, $SD = 1.05$, $\alpha = .81$), Shyness (reversed – slow or inhibited approach and/or discomfort in social situations involving novelty or uncertainty; $M = 3.20$, $SD = .85$, $\alpha = .77$), and Sociability (seeking and taking pleasure in interactions with others; $M = 5.32$, $SD = .95$, $\alpha = .90$).

The second ECBQ higher-order factor is Negative Affect ($M = 2.59$, $SD = .59$, $\alpha = .72$). Negative Affect consists of the following subscales: Discomfort (negative affect related to sensory qualities of stimulation, including intensity, rate or complexity of light, sound, texture; $M = 2.16$, $SD = .83$, $\alpha = .53$), Fear (negative affect, including unease, worry, or nervousness related to anticipated pain or distress and/or potentially threatening situations; startle to sudden events; $M = 2.18$, $SD = .71$, $\alpha = .83$), Frustration (negative affect related to interruption of ongoing tasks or goal blocking; $M = 3.50$, $SD = .91$, $\alpha = .84$), and Sadness (tearfulness or lowered mood related to exposure to personal suffering, disappointment, object loss, loss of approval, or response to other's suffering; $M = 2.54$, $SD = .76$, $\alpha = .72$).

The third ECBQ higher-order factor is Effortful Control ($M = 4.48$, $SD = .46$, $\alpha = .64$, consisting of the following subscales: Attentional Focusing (sustained duration of orienting on an object of attention; resisting distraction; $M = 4.00$, $SD = .80$, $\alpha = .85$), Attentional Shifting (ability to transfer attentional focus from one activity/task to another; $M = 4.46$, $SD = .60$, $\alpha = .68$), Cuddliness (expression of enjoyment in and molding of the body to being held by a caregiver; $M = 4.98$, $SD = .82$, $\alpha = .87$), Low-intensity Pleasure (pleasure or enjoyment related to situations involving low stimulus intensity, rate, complexity, novelty and incongruity; $M = 4.82$, $SD = .82$, $\alpha = .76$), Perceptual Sensitivity (detection of slight, low intensity stimuli from the external environment; $M = 3.75$, $SD = .97$, $\alpha = .76$), Inhibitory Control (capacity to stop, moderate, or refrain from a behavior under instruction; $M = 3.74$, $SD = .94$, $\alpha = .90$), and Soothability (rate of recovery from peak distress, excitement, or general arousal; $M = 5.59$, $SD = .66$, $\alpha = .81$).

Genotyping

Buccal swab samples were acquired from each child. DNA was isolated from the swabs using QuickExtract V1.0 (Epicentre) according to their protocol. We present here specific genotyping procedures for the three genes that are the focus of the current paper: COMT, CHRNA4, and SNAP25.

Genotyping – COMT

The $Val^{108/158}Met$ allele (rs4680) was determined according to Daniels et al. (1996). The reaction components included 0.2 μM of the each primer, 3.0 mM $MgCl_2$, 0.2 mM each dNTP, 1.2% formamide, and 0.05 U/ul Taq polymerase with its 1x reaction buffer $(NH_4)_2SO_4$ (Fermentas). The PCR products were amplified at 94°C 3 min, followed by 40 cycles of 94°C 30 s, 54°C 30 s, 72°C 30 s, and a final incubation at 72°C 3 min. The products were digested with the restriction enzyme NlaIII at 37°C overnight and size-separated on a 4.5% high resolution agarose gel (A4718, Sigma) stained with ethidium bromide. The 169-bp fragment is digested to produce a 114-bp fragment for the valine allele and 96 bp for the methionine allele.

Genotyping – CHRNA4

The CHRNA4 locus was amplified with the primers 5′-ACCAGGG CTGGCCAAAGCCAGG and 5′-GTGCTTTGGTGCTGCGGGTC (Kent et al., 2001). The amplification reaction was similar to that of COMT, with the following changes: 2.5mM $MgCl_2$ in the reaction mix, the denaturing and annealing steps were 20 seconds each, and the annealing temperature was 57°C. The products were digested with the restriction enzyme HhaI (NEB) and the fragments separated on a 4% resolving agarose gel. The amplified product of the C allele of $1860^{C/T}$ polymorphism (rs1044396) is digested twice by HhaI, producing fragments of 152, 105 and 33 bp. The T allele is cut only once, producing 152- and 138-bp fragments.

Genotyping – SNAP25

SNAP25 was amplified using the following primers: 5′-TTCTCCTC CAAATGCTGTCG AND 5′-CCACCGAGGAGAGAAAATG (Barr et al., 2000). The amplification was performed as for COMT, with the addition of 2%

DMSO and an annealing temperature of 55°C. The products were digested with the restriction enzyme DdeI (NEB) and separated on a 3% resolving agarose gel. The C allele of the $1076^{C/T}$ polymorphism (rs1051312) is digested by DdeI to produce 288- and 33-bp fragments from the 261-bp product.

RESULTS

Stability of emotion

Age of the child at Time 1 assessment was significantly correlated with IBQ Positive Affectivity ($r = .29$, $p = .03$) and age at Time 2 was significantly correlated with ECBQ Effortful Control ($r = .28$, $p = .02$). To control for potential age related confounds, partial correlations between the IBQ-R and ECBQ higher order factors were conducted controlling for age at the Time 1 and Time 2 assessments. These partial correlations are presented in Table 2.

Consistent with previous research showing some stability of Positive Affect in infancy, results indicated that Positive Affect as assessed by the IBQ at Time 1 was significantly correlated with Positive Affect as assessed by the ECBQ at Time 2 ($r = .47$, $p = .003$). Our results also showed some stability in Negative Affect ($r = .37$, $p = .02$). There was no evidence that Orienting at Time 1 was significantly related to Effortful Control or other aspects of temperament at Time 2. This finding is also consistent with previous research (Putnam et al., 2008).

Regulation

As expected, orienting as measured from the IBQ was not correlated with Effortful Control as measured from the ECBQ. Orienting was positively

TABLE 2
Correlations-IBQ at Time 1 and ECBQ at Time 2

Time 1-Infant Behavior Questionnaire (IBQ)	*1*	*2*	*3*	*4*	*5*	*6*
1. Positive Affect-Higher Order	1.00					
2. Negative Affect-Higher Order	−.05	1.00				
3. Orienting-Higher Order	.46**	−.33*	1.00			
Time 2-Early Childhood Behavior Questionnaire (ECBQ)						
4. Positive Affect-Higher Order	.47**	−.11	.25	1.00		
5. Negative Affect-Higher Order	.17	.37*	−.08	.07	1.00	
6. Effortful Control-Higher Order	.10	−.06	.26	−.24	−.28	1.00

Note: Correlations were conducted partialling child's age at Time 1 and Time 2.

related to positive affect at Time 1. This was expected because previous work had shown that duration of orienting as reported by caregivers was longer for children higher in smiling and laughter (Rothbart, Derryberry, & Hershey, 2000). Orienting was negatively related to negative affect during infancy. This was expected based upon the role of orienting in soothing (Harman et al., 1997). At Time 2, effortful control was negatively related to both positive and negative affect scores; however, neither correlation was significant. This was unexpected, since effortful control had been found to have a significant negative relation to negative affect in older children and adults. It is possible that toddlers may be at a transition stage, since Orienting and Effortful Control showed both a positive but non significant (see Table 2) relation at the two ages.

COMT and temperament at Time 1

A multivariate analysis of covariance (MANCOVA) was conducted with COMT as the independent variable, age at Time 1 and sex of child as covariates, and the IBQ higher-order factors Positive Affect, Negative Affect, and Orienting as dependent variables. There was evidence of a multivariate main effect of COMT, Pillai's trace = .38, $F(6, 60) = 2.37$, $p = .04$, $\eta_p^2 = .19$. Inspection of the univariate tests showed no evidence of significant effects for either Negative Affect, $F(2, 31) = 1.07$, $p = .35$, $\eta_p^2 = .07$, or Orienting, $F(2, 31) = 2.87$, $p = .07$, $\eta_p^2 = .16$. However, there was evidence of a significant effect for Positive Affect, $F(2, 31) = 5.28$, $p = .01$, $\eta_p^2 = .25$. Follow up tests using paired comparisons indicate that infants who are heterozygous are rated significantly higher in Positive Affect at Time 1 ($M = 5.07$, $SD = .57$), than infants who are either homozygous for valine ($M = 4.37$; $SD = .42$) or methionine ($M = 4.67$, $SD = .46$). The two homozygous groups do not differ from one another significantly.

Examining the IBQ Positive Affect sub-scales individually showed COMT to be significantly related to High Intensity Pleasure, $F(2, 31) = 5.35$, $p = .01$, $\eta_p^2 = .26$, and Smiling and Laughter, $F(2, 31) = 5.00$, $p = .01$, $\eta_p^2 = .24$, such that heterozygous infants were rated higher than both homozygous groups. COMT was also significantly related to Vocal Reactivity, $F(2, 31) = 4.69$, $p = .02$, $\eta_p^2 = .23$. In this case, while the heterozygous infants were rated significantly more positively than the val homozygous infants ($p = .008$), but did not differ significantly from the met homozygous met infants ($p = .07$), there was a trend in that direction. COMT was unrelated to the Approach, Activity Level, and Perceptual Sensitivity subscales of Positive Affect.

COMT and temperament at Time 2

A multivariate analysis of covariance (MANCOVA) was conducted with COMT as the independent variable, age at Time 2 and sex of child as covariates, and the ECBQ higher-order factors Positive Affect, Negative Affect, and Effortful Control as dependent variables. There was no evidence of a multivariate main effect of COMT, Pillai's Trace = .17, $F(6, 80) = 1.24$, $p = .30$, $\eta_p^2 = .09$. Inspection of the univariate tests showed no evidence of significant effects for any of the individual higher order ECBQ factors.

SNAP25 and temperament at Time 1

A multivariate analysis of covariance (MANCOVA) was conducted with SNAP25 as the independent variable, age at Time 1 and sex of child as covariates, and the IBQ higher-order factors Positive Affect, Negative Affect, and Orienting as dependent variables. There was no evidence of a multivariate main effect of SNAP25, $F(3, 30) = 1.95$, $p = .14$, $\eta_p^2 = .16$. Inspection of the univariate tests showed no evidence of significant effects for either Positive Affect, $F(1, 32) = 0.09$, $p = .76$, $\eta_p^2 = .003$, or Orienting, $F(1, 32) = 1.97$, $p = .17$, $\eta_p^2 = .06$. However, there was evidence of a significant effect for Negative Affect, $F(1, 32) = 5.48$, $p = .03$, $\eta_p^2 = .15$, such that infants at Time 1 with one or two copies of the C allele ($M = 2.79$; $SD = .50$) were rated significantly lower in Negative Affect than infants homozygous for the T allele ($M = 3.19$; $SD = .72$).

Examining the IBQ Negative Affect sub-scales individually showed SNAP25 was significantly related to Distress to Limitations, $F(1, 32) = 6.55$, $p = .02$, $\eta_p^2 = .17$, such that infants at Time1 with one or two copies of the C allele ($M = 3.22$; $SD = .65$) were rated significantly lower in Distress to Limitations than infants homozygous for the T allele ($M = 3.81$; $SD = .85$).

SNAP25 and temperament at Time 2

A multivariate analysis of covariance (MANCOVA) was conducted with SNAP25 as the independent variable, age at Time 2 and sex of child as covariates, and the ECBQ higher-order factors Positive Affect, Negative Affect, and Effortful Control as dependent variables. There was no evidence of a multivariate main effect of SNAP25, Pillai's Trace = .12, $F(3, 40) = 1.76$, $p = .17$, $\eta_p^2 = .12$. Inspection of the univariate tests showed no evidence of significant effects for either Positive Affect, $F(1, 42) = 0.30$, $p = .59$, $\eta_p^2 = .007$, or Effortful Control, $F(1, 42) = 0.03$, $p = .87$, $\eta_p^2 = .001$. However, there was evidence of a significant effect for Negative Affect, $F(1, 42) = 4.42$, $p = .04$, $\eta_p^2 = .10$, such that infants at Time 2 with a one or two copies of the

C allele ($M = 2.41$, $SD = .55$) were rated significantly lower in Negative Affect than infants homozygous for the T allele ($M = 2.73$, $SD = .53$).

Examining the ECBQ Negative Affect sub-scales individually showed the SNAP25 was significantly related to Discomfort, $F(1, 42) = 4.56$, $p = .04$, $\eta_p^2 = .10$, such that infants at Time 1 with one or two copies of the C allele ($M = 1.82$; $SD = .70$) were rated significantly lower in Discomfort than infants homozygous for the T allele ($M = 2.30$; $SD = .76$). There was no evidence that SNAP25 was significantly related to the Fear, Frustration, or Sadness subscales.

CHRNA4 and temperament at Time 1

A multivariate analysis of covariance (MANCOVA) was conducted with CHRNA4 as the independent variable, age at Time 1 and sex of child as covariates, and the IBQ higher-order factors Positive Affect, Negative Affect, and Orienting as dependent variables. There was no evidence of a multivariate main effect of CHRNA4, Pillai's Trace = .11, $F(6, 60) = 0.56$, $p = .76$, $\eta_p^2 = .05$. Inspection of the univariate tests showed no evidence of significant effects for any of the individual higher order IBQ factors.

CHRNA4 and temperament at Time 2

A multivariate analysis of covariance (MANCOVA) was conducted with CHRNA4 as the independent variable, age at Time 2 and sex of child as covariates, and the ECBQ higher-order factors Positive Affect, Negative Affect, and Effortful Control as dependent variables. There was no evidence of a multivariate main effect of CHRNA4, Pillai's Trace = .27, $F(6, 80) = 2.12$, $p = .06$, $\eta_p^2 = .14$. Inspection of the univariate tests showed no evidence of significant effects for either Positive Affect, $F(2, 41) = 0.15$, $p = .86$, $\eta_p^2 = .007$, or Negative Affect, $F(2, 41) = 0.14$, $p = .87$, $\eta_p^2 = .007$. However, there was evidence of a significant effect for Effortful Control, $F(2, 41) = 6.15$, $p = .005$, $\eta_p^2 = .23$, such that T/T homozygous toddlers were rated significantly lower in Effortful Control at Time 2 ($M = 4.13$, $SD = .32$) than both heterozygous infants ($M = 4.36$, $SD = .43$) and C/C homozygous infants ($M = 4.64$, $SD = .35$). T/C and C/C infants did not differ significantly from one another ($p = .08$).

Examining the ECBQ Effortful Control sub-scales individually showed CHRNA4 was significantly related to Cuddliness, $F(2, 41) = 3.90$, $p = .03$, $\eta_p^2 = .16$, such that T/T homozygous toddlers were rated significantly lower in Cuddliness at Time 2 ($M = 4.37$, $SD = 1.14$) than C/C homozygous infants ($M = 5.29$, $SD = .50$), while the heterozygous infants ($M = 4.85$, $SD = .80$) did not differ significantly from either homozygous group. CHRNA4 was not significantly related to any other Effortful Control subscale; however, there was a pattern of marginal relations with Attentional Focusing ($p = .09$), Attentional Shifting ($p = .08$), and Perceptual Sensitivity ($p = .08$), suggesting that the link between CHRNA4 and Effortful Control is not solely related to differences in cuddliness.

Candidate genes and temperament at Time 1 and Time 2

Estimated marginal means, standard errors, and 95% confidence intervals for temperament outcomes at Time 1 and Time 2 as a function of genotype are reported in Table 3. Estimated marginal means include age in days at time of assessment and sex of child as covariates. Caution is warranted in interpreting null effects due to small sample size and uneven cell distributions.

DISCUSSION

Rothbart and Derryberry (1981) proposed a distinction between reactive and self-regulatory aspects of temperament. They did so based on their finding of separate dimensions of temperament scales involved with infant emotion and behavior from those relating to the ability to regulate that emotion or behavior. Moreover, they argued for developmental changes in which the regulatory aspects of temperament might arise later in development than the reactive aspects of temperament.

Our current longitudinal study has begun to confirm, revise and extend that analysis. We find a negative correlation at 6–7 months between parental reports of orienting of attention and negative affect, and orienting correlates positively with reports of positive affect. By 18–20 months orienting is no longer related to affect, but effortful control shows modest non significant negative correlations with both positive and negative affect. There is substantial evidence that for children, adolescents and adults effortful control shows a systematic negative correlation with negative affect in western countries, but Ahadi, Rothbart, and Ye (1993) found that in China effortful control had a negative correlation with positive affect in children. They argued that culture shaped the direction of the interaction of effortful control and emotion. We will be obtaining results from 4 year olds in our longitudinal study and we expect to find significant negative correlations between effortful control and negative affect.

The results of our longitudinal study suggest that early in life, orienting serves as a regulatory system and at this age it both reduces negative affect and increases positive affect. However, imaging studies of error detection make it appear likely that the executive attention system is already present by 7 months (Berger, Tzur, & Posner, 2006) and that it becomes dominant as a control mechanism later in life (Rothbart & Rueda, 2005). Thus, during infancy both orienting and executive networks serve parallel regulatory functions.

The nicotinic cholinergic receptor modulates the release of dopamine in the mesolimbic system (Pidoplichko, DeBiasi, Williams, & Dani, 1997). Activation of the α4 subunit of the nicotinic cholinergic receptor is sufficient for nicotine-induced reward, tolerance and sensitization in mice (Tapper et al., 2004). Polymorphisms in the CHRNA4 gene have been associated with nicotine dependence in humans (Hutchinson et al., 2007) and with cognitive performance (Rigbi et al., 2008). The CHRNA4 C1545T polymorphism (rs1044396) has been associated with variation in performance of visuospatial attention (Greenwood, Fossella, & Parasuraman, 2005; Greenwood et al., 2008; Parasuraman, Greenwood, Kumar, & Fossella, 2005) and differences in brain activity when performing visual attention tasks (Winterer et al., 2007).

Since visuospatial attention requires orienting, we expected the CHRNA4 polymorphism to influence orienting in our subjects and thought it might also influence higher order attention. Our genetic findings provide support for the parallel model of regulation discussed above. At 6–7 months CHRNA4 appears to relate to aspects of anticipatory looking. We have regarded anticipatory, but not reactive looking as a very early arising indicant of executive attention. However, this form of looking may also relate to the orienting network. At two years the main influence of this gene appears to be on effortful control, which depends upon executive attention. In adults CHRNA4 seems to be related to tasks that clearly involve the orienting network (Parasuraman et al., 2005), but these task may involve executive attention as well.

During infancy the T/T allele of CHRNA4 is related to better performance in anticipatory looking, but at Time 2 the C/C homozygotes have the highest scores on effortful control. Since we regard anticipatory looking as an early measure of executive attention and effortful control also relates to executive attention this finding is puzzling. There is clearly cholinergic input both to the executive attention network and the orienting network. The surprising finding that alleles associated with higher attention during infancy (as measured by anticipation) have lower parent reported effortful control at Time 2 may indicate that infants who control emotions through orienting are slower to transition to emotional control via the executive network. Since our measurement at 18–20 months occurs well before regulation by executive attention is complete, infants with strong orienting may not have made the transition. Perhaps our tests on these same children at age 4, when we will use both anticipatory looking and the attention network test (Fan, McCandliss, Sommer, Raz, & Posner, 2002), will further explicate the relation of CHRNA4 to orienting and executive attention.

COMT plays an important role in dopamine metabolism by modulating extracellular levels of dopamine. The functional Val/Met polymorphism of COMT has a measurable effect on COMT enzyme activity, where the Val allele can degrade extracellular dopamine more quickly than the less enzymatically active Met allele (Lachman et al., 1996). The Met allele has been associated with anxiety and negative mood states (Drabant et al., 2006; Montag et al., 2008; Smolka et al., 2005; Stein, Fallin, Schork, & Gelernter, 2005), affective disorders (Eley et al., 2003; Karayiorgou et al., 1997; Olsson et al., 2005), and decreased novelty seeking (Reuter & Hennig, 2005; Tsai, Hong, Yu, & Chen, 2004). A finding from our current longitudinal study is that the COMT gene, which has consistently been shown to be related to executive attention in adults and older children, is also related to aspects of executive attention in toddlers (Voelker, Sheese, Rothbart, & Posner, in press). We found that haplotypes of the COMT gene influence both anticipatory looking and nesting cup activity at 18 months, which are both tasks thought to be related to executive attention. At 7 months COMT is also related to positive affect as reported by parents. The finding of a relation of COMT to positive affect together with the influence of this gene on executive attention at 18 months could provide a genetic link between reactive emotion and emotional regulation during early development. However, it is also possible that COMT's relation to positive affect in infancy is mediated by regulatory aspects of executive attention. It is not unlikely that the earliest form of executive attention is regulation of emotion and this may occur in parallel with regulation by orienting. The evidence for this ideas is mixed in our current study in that positive affect in infancy was unrelated to effortful control (see Table 2), but two other studies did show such a connection (Komsi et al., 2006; Putnam et al., 2008).

SNAP-25 is a core component of the SNAP complex, which mediates presynaptic vesicle trafficking. SNAP-25 participates directly in the release of neurotransmitters. Mice heterozygous for a SNAP-25 deficiency are defective in neuronal dopamine signaling and exhibit symptoms of hyperactivity (Hess, Jinnah, Kozak, & Wilson, 1992) and transgenic rescue of SNAP-25 restores dopamine-modulated synaptic transmission (Hess, Collins, & Wilson, 1996). SNAP-25 is a candidate gene for ADHD in humans (Mill et al., 2005), and the MnlI (rs3746544) and DdeI (rs1051312) polymorphisms have been linked to ADHD (Barr et al., 2000; Brophy, Hawi, Kirley, Fitzgerald, & Gill, 2002; Kim et al., 2007). The DdeI polymorphism was also linked to cognitive performance with verbal memory, executive functions, and a general cognitive index in schizophrenia patients (Spellmann et al., 2008). Because of its role modulating dopamine signaling and its association with cognitive abilities, SNAP-25 is a good candidate for

TABLE 3
Estimated marginal means, standard errors, and 95% confidence intervals for temperament and Time 1 and Time 2

Genotype	Temperament at Time 1 (IBQ)													Temperament at Time 2 (ECBQ)													
		Positive Affect				Negative Affect				Orienting					Positive Affect				Negative Affect				Effortful Control			
Groupings	N	M	SE	CI-L	CI-U	M	SE	CI-L	CI-U	M	SE	CI-L	CI-U	N	M	SE	CI-L	CI-U	M	SE	CI-L	CI-U	M	SE	CI-L	CI-U
COMT (rs4680)																										
G/G	7	4.41	0.19	4.02	4.81	3.24	0.25	2.73	3.76	4.61	0.19	4.23	5.00	8	4.37	0.13	4.11	4.63	2.87	0.20	2.47	3.26	4.24	0.15	3.94	4.54
A/G	19	5.08	0.12	4.84	5.32	2.88	0.15	2.57	3.20	5.13	0.12	4.90	5.37	25	4.32	0.07	4.17	4.46	2.46	0.11	2.24	2.68	4.38	0.08	4.22	4.55
A/A	10	4.62	0.17	4.29	4.96	3.19	0.22	2.75	3.63	4.92	0.16	4.60	5.25	13	4.44	0.10	4.24	4.64	2.74	0.15	2.44	3.05	4.45	0.11	4.22	4.68
CHRNA4 (rs1044396)																										
C/C	11	4.88	0.18	4.50	5.25	3.30	0.21	2.88	3.72	4.91	0.17	4.57	5.25	12	4.31	0.11	4.10	4.53	2.55	0.17	2.21	2.89	4.63	0.11	4.41	4.85
T/C	19	4.81	0.14	4.53	5.09	2.96	0.15	2.64	3.27	5.05	0.12	4.80	5.30	22	4.36	0.08	4.21	4.52	2.65	0.12	2.41	2.90	4.39	0.08	4.23	4.55
T/T	6	4.77	0.24	4.28	5.26	2.82	0.27	2.27	3.37	4.86	0.22	4.41	5.30	12	4.40	0.11	4.18	4.61	2.59	0.17	2.25	2.92	4.10	0.11	3.88	4.31
DBH (rs1108580)																										
G/G	13	5.02	0.17	4.68	5.36	3.06	0.19	2.68	3.44	5.23	0.14	4.94	5.51	16	4.51	0.09	4.33	4.69	2.49	0.14	2.21	2.78	4.43	0.10	4.22	4.64
A/G	17	4.72	0.14	4.44	5.00	3.19	0.16	2.87	3.51	4.75	0.12	4.51	4.99	23	4.28	0.07	4.14	4.43	2.71	0.12	2.47	2.95	4.30	0.09	4.12	4.47
A/A	6	4.69	0.25	4.18	5.20	2.56	0.28	1.98	3.14	5.07	0.21	4.64	5.50	7	4.28	0.13	4.01	4.55	2.54	0.22	2.11	2.98	4.52	0.16	4.20	4.83
MAOA (30bp VNTR)																										
3/3-repeat	15	4.82	0.16	4.50	5.15	2.95	0.19	2.57	3.33	5.03	0.14	4.73	5.32	14	4.42	0.10	4.23	4.62	2.68	0.16	2.36	3.00	4.32	0.12	4.08	4.56
3/4-repeat	6	4.85	0.29	4.26	5.44	3.25	0.33	2.57	3.94	5.06	0.26	4.53	5.59	5	4.30	0.18	3.94	4.67	2.65	0.29	2.06	3.24	4.30	0.22	3.86	4.73
4/4-repeat	13	4.79	0.17	4.44	5.14	3.03	0.20	2.63	3.44	4.87	0.15	4.56	5.18	25	4.36	0.07	4.22	4.50	2.59	0.11	2.36	2.82	4.41	0.09	4.24	4.58
SNAP25 MnlI (rs3746544)																										
G/G	9	4.97	0.19	4.57	5.37	3.15	0.24	2.66	3.63	5.07	0.18	4.69	5.44	7	4.45	0.14	4.17	4.72	3.08	0.20	2.67	3.49	4.22	0.16	3.90	4.53
G/T	14	4.97	0.15	4.66	5.28	2.95	0.19	2.57	3.33	5.04	0.14	4.75	5.34	23	4.27	0.07	4.12	4.41	2.53	0.11	2.30	2.75	4.37	0.09	4.20	4.54
T/T	13	4.57	0.16	4.23	4.90	3.06	0.20	2.65	3.46	4.84	0.16	4.52	5.15	16	4.46	0.09	4.28	4.64	2.52	0.13	2.25	2.79	4.45	0.10	4.24	4.66

TABLE 3 (*Continued*)

Genotype Groupings		Temperament at Time 1 (IBQ)													Temperament at Time 2 (ECBQ)											
		Positive Affect				Negative Affect				Orienting					Positive Affect				Negative Affect				Effortful Control			
	N	*M*	*SE*	*CI-L*	*CI-U*	*M*	*SE*	*CI-L*	*CI-U*	*M*	*SE*	*CI-L*	*CI-U*	*N*	*M*	*SE*	*CI-L*	*CI-U*	*M*	*SE*	*CI-L*	*CI-U*	*M*	*SE*	*CI-L*	*CI-U*
TPH2 (rs4570625)																										
T absent	20	4.82	0.13	4.55	5.08	3.05	0.15	2.74	3.36	4.89	0.12	4.65	5.13	25	4.30	0.07	4.16	4.45	2.72	0.11	2.50	2.95	4.31	0.08	4.15	4.48
T present	16	4.83	0.15	4.54	5.13	3.02	0.17	2.68	3.36	5.08	0.13	4.81	5.34	21	4.43	0.08	4.27	4.58	2.47	0.12	2.23	2.72	4.45	0.09	4.27	4.63
5HTT (SLC6A4) (5HTTLPR 44bp in/del)																										
short (44bp del) present	26	4.82	0.11	4.59	5.06	3.09	0.13	2.82	3.36	4.94	0.10	4.73	5.15	35	4.35	0.06	4.23	4.48	2.63	0.10	2.44	2.83	4.35	0.07	4.21	4.48
short absent	10	4.82	0.19	4.45	5.20	2.91	0.21	2.47	3.34	5.08	0.17	4.73	5.42	11	4.39	0.11	4.16	4.61	2.53	0.17	2.19	2.88	4.47	0.12	4.22	4.72
APOE (rs429358 and rs7412)																										
TT or CT present	26	4.82	0.12	4.58	5.06	2.89	0.13	2.63	3.15	4.99	0.11	4.78	5.21	32	4.34	0.06	4.22	4.47	2.59	0.10	2.39	2.80	4.40	0.07	4.25	4.54
TT or CT absent	10	4.84	0.19	4.45	5.23	3.42	0.21	3.00	3.85	4.92	0.18	4.57	5.28	14	4.40	0.10	4.20	4.59	2.64	0.15	2.34	2.95	4.32	0.11	4.10	4.55
BDNF (rs6265)																										
A(met) present	12	4.85	0.17	4.50	5.21	3.03	0.20	2.62	3.44	4.99	0.16	4.67	5.32	15	4.33	0.09	4.14	4.51	2.49	0.14	2.20	2.78	4.40	0.11	4.19	4.61
A absent	24	4.81	0.12	4.56	5.06	3.04	0.14	2.76	3.33	4.97	0.11	4.74	5.19	31	4.38	0.06	4.25	4.51	2.67	0.10	2.47	2.87	4.36	0.07	4.22	4.51
DAT1(SLC6A3) (rs28363170)																										
9-repeat present	26	4.82	0.11	4.59	5.06	3.09	0.13	2.82	3.36	4.94	0.10	4.73	5.15	24	4.33	0.07	4.19	4.48	2.73	0.11	2.51	2.96	4.29	0.08	4.12	4.45
9-repeat absent	10	4.82	0.19	4.45	5.20	2.91	0.21	2.47	3.34	5.08	0.17	4.73	5.42	22	4.39	0.08	4.24	4.54	2.47	0.12	2.24	2.71	4.47	0.09	4.30	4.65

TABLE 3 (*Continued*)

Genotype Groupings	*N*	*M*	*SE*	*CI-L*	*CI-U*	*M*	*SE*	*CI-L*	*CI-U*	*M*	*SE*	*CI-L*	*CI-U*	*N*	*M*	*SE*	*CI-L*	*CI-U*	*M*	*SE*	*CI-L*	*CI-U*	*M*	*SE*	*CI-L*	*CI-U*
	Temperament at Time 1 (IBQ)													*Temperament at Time 2 (ECBQ)*												
		Positive Affect				*Negative Affect*				*Orienting*					*Positive Affect*				*Negative Affect*				*Effortful Control*			
DRD4, 7-repeat (48bp VNTR)																										
7-repeat present	24	4.83	0.12	4.59	5.07	3.04	0.14	2.76	3.33	4.96	0.11	4.74	5.18	32	4.33	0.06	4.21	4.46	2.61	0.10	2.41	2.81	4.40	0.07	4.26	4.55
7-repeat absent	12	4.81	0.17	4.47	5.16	3.02	0.20	2.62	3.43	5.00	0.15	4.69	5.31	14	4.42	0.10	4.22	4.61	2.61	0.15	2.30	2.91	4.31	0.11	4.09	4.53
DRD4, 4-repeat (48bp VNTR)																										
non 4-repeat homozygote	20	4.83	0.13	4.57	5.10	2.92	0.15	2.61	3.22	5.04	0.12	4.80	5.28	24	4.37	0.08	4.22	4.53	2.63	0.12	2.39	2.87	4.33	0.09	4.16	4.50
4-repeat homozygote	16	4.81	0.15	4.51	5.11	3.19	0.17	2.85	3.53	4.89	0.13	4.62	5.16	22	4.35	0.08	4.19	4.51	2.59	0.12	2.34	2.84	4.43	0.09	4.25	4.61
SNAP25 DdeI (rs1051312)																										
C present	14	4.86	0.16	4.54	5.19	2.72	0.17	2.37	3.07	5.13	0.14	4.84	5.42	18	4.32	0.09	4.15	4.50	2.40	0.13	2.14	2.66	4.39	0.10	4.19	4.58
C absent	24	4.80	0.13	4.54	5.06	3.24	0.14	2.97	3.51	4.87	0.11	4.65	5.10	28	4.38	0.07	4.25	4.52	2.74	0.10	2.54	2.95	4.37	0.08	4.21	4.52

Note. CI-L and CI-U indicate the lower and upper bounds of the 95% confidence interval, respectively. Estimated marginal means are calculated with age in days at time of assessment and sex of child included as covariates.

behavioral studies. SNAP25 has been associated with a variety of mental disorders including bipolar disorder, schizophrenia and Attention Deficit Hyperactivity Disorder. Since these disorders often have in common negative affect it is not surprising that in our normative sample we find an association as early as 7 months between this gene and negative affect.

A number of the candidate genes were expected to show relations to emotion and self-regulation that were not found in our data. For example, in adults, genes in the dopamine (COMT, DRD4, DAT1, DBH) and serotonin (TRP2, 5HTT) pathways have been found to be related to executive attention as measured by the attention network test and other conflict related tasks. One of the most powerful relations between negative affect has been reported for the serotonin transporter gene 5HTT in children and adults. However, our work with infants does not show any relation of 5HTT to negative affect. We did find effects of the DRD4 7-repeat allele on a number of parent reported temperament variables closely related to symptoms of ADHD, but since this effect interacted with parental quality we have reported these findings in a separate paper (Sheese et al., 2007).

We can evaluate our four hypotheses in light of these results. There does seem a separation of genes supporting positive and negative affect in infancy (Hypothesis 1). Moreover both positive and negative affect are related to orienting as suggested by Hypothesis 2. The influence of effortful control on negative emotionality as suggested by Hypothesis 3 was not significant at age 2, although we will examine the relationship again at age 4. The COMT finding shows that the same gene can be related both to reactive emotion (positive affect in infancy) and self-regulation in violation of Hypothesis 4. Whether this is a general violation of the idea that these networks have separate genetic origins or whether it will prove merely one exception to the more general separation between networks remains to be seen.

Because of the relatively small N of our study and the many comparisons involved all these results must be taken as tentative and awaiting further confirmation. However, our study does suggest the possibility of using behavioral and genetic studies to test specific ideas of what is changing in development.

Overall these data support the important role of executive attention as a regulatory mechanism by 18–20 months as well as the role of orienting as a control mechanism at 6–7 months. It seems likely that the parallel activity of these two networks begins in infancy and continues to adulthood. The strong tendency for adults to look away as a self-regulatory strategy suggests a continued role for orienting even for adults. We hope our future longitudinal results and additional studies will provide information on the coordination between these regulatory networks.

REFERENCES

Ahadi, S. A., Rothbart, M. K., & Ye, R. (1993). Children's temperament in the U.S. and China: Similarities and differences. *European Journal of Personality, 7*, 359–378.

Auerbach, J., Geller, V., Lezer, S., Shinwell, E., Belmaker, R. H., Levine, J., & Ebstein, R. (1999). Dopamine D4 receptor (D4DR) and serotonin transporter promoter (5-HTTLPR) polymorphisms in the determination of temperament in 2-month-old infants. *Molecular Psychiatry, 4*, 369–373.

Auerbach, J. G., Faroy, M., Ebstein, R., Kahana, M., & Levine, J. (2001). The association of the dopamine D4 receptor gene (DRD4) and the serotonin transporter promoter gene (5 HTTLPR) with temperament in 12-month-old infants. *Journal of Child Psychology and Psychiatry and Allied Disciplines, 42*, 777–783.

Bakermans-Kranenburg, M. J., & van IJzendoorn, M. H. (2006). Gene-environment interaction of the dopamine D4 receptor (DRD4) and observed maternal insensitivity predicting externalizing behavior in preschoolers. *Developmental Psychobiology, 48*, 406–409.

Bakermans-Kranenburg, M. J., Van IJzendoorn, M. H., Pijlman, F. T., Mesman, J., & Juffer, F. (2008). Experimental evidence for differential susceptibility: Dopamine D4 Receptor Polymorphism (DRD4 VNTR) moderates intervention effects on toddlers' externalizing behavior in a randomized controlled trial. *Developmental Psychology, 44*, 293–300.

Barr, C. L., Feng, Y., Wigg, K., Bloom, S., Roberts, W., Malone, M., Schachar, R., Tannock, R., & Kennedy, J. L. (2000). Identification of DNA variants in the SNAP-25 gene and linkage study of these polymorphisms and attention-deficit hyperactivity disorder. *Molecular Psychiatry, 5*, 405–409.

Berger, A., Tzur, G., & Posner, M. I. (2006). Infant brains detect arithmetic errors. *Proceedings of the National Academy of Science of the United States of America, 103*, 12649–12653.

Brophy, K., Hawi, Z., Kirley, A., Fitzgerald, M., & Gill, M. (2002). Synaptosomal-associated protein 25 (SNAP-25) and attention deficit hyperactivity disorder (ADHD): Evidence of linkage and association in the Irish population. *Molecular Psychiatry, 7*, 913–917.

Caspi, A., McClay, J., Moffitt, T. E., Mill, J., Martin, J., Craig, I. W., Taylor, A., & Poulton, R. (2002). Role of genotype in the cycle of violence in maltreated children. *Science, 297*, 851–854.

Daniels, J. K., Williams, N. M., Williams, J., Jones, L. A., Cardno, A. G., Murphy, K. C., Spurlock, G., Riley, B., Scambler, P., Asherson, P., McGuffin, P., & Owen, M. J. (1996). No evidence for allelic association between schizophrenia and a polymorphism determining high or low catechol O-methyltransferase activity. *American Journal of Psychiatry, 153*, 268–270.

Deckert, J., Catalano, M., Heils, A., Di Bella, D., Friess, F., Politi, E., Franke, P., Nöthen, M. M., Maier, W., Bellodi, L., & Lesch, K. P. (1997). Functional promoter polymorphism of the human serotonin transporter: Lack of association with panic disorder. *Psychiatric Genetics, 7*, 45–47.

De Luca, A., Rizzardi, M., Torrente, I., Alessandroni, R., Salvioli, G. P., Filograsso, N., Dallapiccola, B., & Novelli, G. (2001). Dopamine D4 receptor (DRD4) polymorphism and adaptability trait during infancy: A longitudinal study in 1- to 5-month-old neonates. *Neurogenetics, 3*, 79–82.

De Luca, A., Rizzardi, M., Buccino, A., Alessandroni, R., Salvioli, G. P., Filograsso, N., Novelli, G., & Dallapiccola, B. (2003). Association of dopamine D4 receptor (DRD4) exon III repeat polymorphism with temperament in 3-year-old infants. *Neurogenetics, 4*, 207–212.

Drabant, E. M., Hariri, A. R., Meyer-Lindenberg, A., Munoz, K. E., Mattay, V. S., Kolachana, B. S., Egan, M. F., & Weinberger, D. R. (2006). Catechol O-methyltransferase val158met geno type and neural mechanisms related to affective arousal and regulation. *Archives of General Psychiatry, 63*, 1396–1406.

Eley, T. C., Tahir, E., Angleitner, A., Harriss, K., McClay, J., Plomin, R., Riemann, R., Spinath, F., & Craig, I. (2003). Association analysis of MAOA and COMT with neuroticism assessed by peers. *American Journal of Medical Genetics part B: Neuropsychiatric Genetics, 120B*, 90–96.

Esposito, E., DiMatteo, V., & Di Giovanni, G. (2008). Serotonin–dopamine interaction: An overview. *Progress in Brain Research, 172*, 3–6.

Fan, J., McCandliss, B. D., Sommer, T., Raz, A., & Posner, M. I. (2002). Testing the efficiency and independence of attentional networks. *Journal of Cognitive Neuroscience, 14*, 340–347.

Forbes, E. E., Brown, S. M., Kimak, M., Ferrell, R. E., Manuck, S. B., & Hariri, A. R. (2009). Genetic variation in components of dopamine neurotransmission impacts ventral striatal reactivity associated with impulsivity. *Molecular Psychiatry, 14*, 60–70.

Gartstein, M. A., & Rothbart, M. K. (2003). Studying infant temperament via the Revised Infant Behavior Questionnaire. *Infant Behavior and Development, 26*, 64–86.

Gillihan, S. J., Farah, M. J., Sankoorikal, G. M., Breland, J., & Brodkin, E. S. (2007). Association between serotonin transporter genotype and extraversion. *Psychiatric Genetics, 17*, 351–354.

Greenwood, P. M., Fossella, J. A., & Parasuraman, R. (2005). Specificity of the effect of a nicotinic receptor polymorphism on individual differences in visuospatial attention. *Journal of Cognitive Neuroscience, 17*, 1611–1620.

Greenwood, P. M., Sundararajan, R., Lin, M. K., Kumar, R., Fryxell, K. J., & Parasuraman, R. (2008). Both a nicotinic SNP and a noradrenergic SNP modulate working memory performance when attention is manipulated. *Journal of Cognitive Neuroscience, Epub ahead of print.*

Harman, C., Rothbart, M. K., & Posner, M. I. (1997). Distress and attention interactions in early infancy. *Motivation and Emotion, 21*, 27–43.

Hess, E. J., Jinnah, H. A., Kozak, C. A., & Wilson, M. C. (1992). Spontaneous locomotor hyperactivity in a mouse mutant with a deletion including the SNAP gene on chromosome 2. *Journal of Neuroscience, 12*, 2865–2874.

Hess, E. J., Collins, K. A., & Wilson, M. C. (1996). Mouse model of hyperkinesis implicates SNAP-25 in behavioral regulation. *Journal of Neuroscience, 16*, 3104–3111.

Hutchinson, K. E., Allen, D. L., Filbey, F. M., Jepson, C., Lerman, C., Benowitz, N. L., Stitzel, J., Bryan, A., McGeary, J., & Haughey, H. M. (2007). CHRNA4 and tobacco dependence: From gene regulation to treatment outcome. *Archives of General Psychiatry, 64*, 1078–1086.

Karayiorgou, M., Altemus, M., Galke, B. L., Goldman, D., Murphy, D. L., Ott, J., & Gogos, J. A. (1997). Genotype determining low catechol-O-methyltransferase activity as a risk factor for obsessive-compulsive disorder. *Proceedings of the National Academy of Sciences of the United States of America, 94*, 4572–4575.

Kent, L., Middle, F., Hawi, Z., Fitzgerald, M., Gill, M., Feehan, C., & Craddock, N. (2001). Nicotinic acetylcholine receptor (4 subunit gene polymorphism and attention deficit hyperactivity disorder. *Psychiatric Genetics, 11*, 37–40.

Kim, J. W., Biederman, J., Arbeitman, L., Fagerness, J., Doyle, A. E., Petty, C., Perlis, R. H., Purcell, S., Smoller, J. W., Faraone, S. V., & Sklar, P. (2007). Investigation of variation in SNAP-25 and ADHD and relationship to co-morbid major depressive disorder. *American Journal of Medical Genetics Part B: Neuropsychiatric Genetics, 144B*, 781–790.

Komsi, N., Räikkönen, K., Pesonen, A. K., Heinonen, K., Keskivaara, P., Järvenpää, A. L., & Strandberg, T. E. (2006). Continuity of temperament from infancy to middle childhood. *Infant Behavior & Development, 29*, 494–508.

Lachman, H. M., Papolos, D. F., Saito, T., Yu, Y. M., Szumlanski, C. L., & Weinshilboum, R. M. (1996). Human catechol-O-methyltransferase pharmacogenetics: Description of a functional polymorphism and its potential application to neuropsychiatric disorders. *Pharmacogenetics, 6*(3), 243–250.

Lakatos, K., Nemoda, Z., Birkas, E., Ronai, Z., Kovacs, E., Ney, K., Toth, I., Sasvari-Szekely, M., & Gervai, J. (2003). Association of D4 dopamine receptor gene and serotonin transporter promoter polymorphisms with infants' response to novelty. *Molecular Psychiatry, 8*, 90–97.

Mill, J., Xu, X., Ronald, A., Curran, S., Price, T., Knight, J., Craig, I., Sham, P., Plomin, R., & Asherson, P. (2005). Quantitative trait locus analysis of candidate gene alleles associated with attention deficit hyperactivity disorder (ADHD) in five genes: DRD4, DAT1, DRD5, SNAP-25, and 5HT1B. *American Journal of Medical Genetics B: Neuropsychiatric Genetics, 133B*, 68–73.

Montag, C., Buckholtz, J. W., Hartmann, P., Merz, M., Burk, C., Hennig, J., & Reuter, M. (2008). COMT genetic variation affects fear processing: psychophysiological evidence. *Behavioral Neuroscience, 122*, 901–909.

Nigg, J., Friderici, K., Park, L., & Zucker, R. A. (2007). Genotype and neuropsychological response inhibition as resilience promoters for attention deficit/hyperactivity disorder, oppositional defiant disorder and conduct disorder under conditions of psychosocial adversity. *Development and Psychopathology, 19*, 767–786.

Olsson, C. A., Anney, R. J., Lotfi-Miri, M., Byrnes, G. B., Williamson, R., & Patton, G. C. (2005). Association between the COMT Val158Met polymorphism and propensity to anxiety in an Australian population- based longitudinal study of adolescent health. *Psychiatric Genetics, 15*, 109–115.

Parasuraman, R., Greenwood, P. M., Kumar, R., & Fossella, J. (2005). Beyond heritability – neurotransmitter genes differentially modulate visuospatial attention and working memory. *Psychological Science, 16*, 200–207.

Plomin, R., Kennedy, J. K. J., & Craig, I. W. (2006). The quest for quantitative trait loci associated with intelligence. *Intelligence, 34*, 513–526.

Posner, M. I., & Rothbart, M. K. (2000). Developing mechanisms of self-regulation. *Development and Psychopathology, 12*, 427–441.

Posner, M. I., & Rothbart, M. K. (2007). *Educating the human brain*. Washington, DC: APA Books.

Posner, M. I., Rothbart, M. K., & Sheese, B. E. (2007). Attention genes. *Developmental Science, 10*, 24–29.

Pidoplichko, V. I., DeBiasi, M., Williams, J. T., & Dani, J. A. (1997). Nicotine activates and desensitizes midbrain dopamine neurons. *Nature, 390*, 401–404.

Putnam, S. P., Gartstein, M. A., & Rothbart, M. K. (2006). Measurement of fine-grained assessment of toddler temperament: The Early Childhood Behavior Questionnaire. *Infant Behavior and Development, 29*, 386–401.

Putnam, S. P., Rothbart, M. K., & Gartstein, M. A. (2008). Homotypic and heterotypic continuity of fine-grained temperament during infancy, toddlerhood, and early childhood. *Infant and Child Development, 17*, 387–405.

Rigbi, A., Kanyas, K., Yakir, A., Greenbaum, L., Pollak, Y., Ben-Asher, E., Lancet, D., Kertzman, S., & Lerer, B. (2008). Why do young women smoke? V. Role of direct and interactive effects of nicotinic cholinergic receptor gene variation on neurocognitive function. Genes, Brain. *and Behavior, 7*, 164–172.

Rothbart, M. K. (2007). Temperament, development, and personality. *Current Directions in Psychological Science, 16*, 207–212.

Rothbart, M. K., & Bates, J. E. (2006). Temperament in children's development. In W. Damon & R. Lerner (Book Eds.) & N. Eisenberg (Vol. Ed.), *Handbook of child psychology: Vol. 3. Social, emotional, and personality development* (6th ed., pp. 99–166). New York: Wiley.

Rothbart, M. K., & Derryberry, D. (1981). Development of individual differences in temperament. In M. E. Lamb & A. L. Brown (Eds.), *Advances in developmental psychology* (Vol. 1, pp. 37–86). Hillsdale, NJ: Erlbaum.

Rothbart, M. K., & Posner, M. I. (2006). Temperament, attention, and developmental psychopathology. In D. Cicchetti & D. J. Cohen (Eds.), *Handbook of developmental psychopathology* (Vol. 2, pp. 465–501). New York: Wiley Press.

Rothbart, M. K., & Rueda, M. R. (2005). The development of effortful control. In U. Mayr, E. Awh, & S. W. Keele (Eds.), *Developing individuality in the human brain: A tribute to Michael I. Posner* (pp. 167–188). Washington, DC: American Psychological Association.

Rothbart, M. K., Derryberry, D., & Hershey, K. (2000). Stability of temperament in childhood: Laboratory infant assessment to parent report at seven years. In V. J. Molfese & D. L. Molfese (Eds.), *Temperament and personality development across the life span* (pp. 85–119). Hillsdale, NJ: Erlbaum.

Reuter, M., & Hennig, J. (2005). Association of the functional catechol-O-methyltransferase Val158Met polymorphism with the personality trait of extraversion. *Neuroreport, 16*, 1135–1138.

Sabol, S. Z., Hu, S., & Hamer, D. (1998). A functional polymorphism in the monoamine oxidase A gene promoter. *Human Genetics, 103*, 273–279.

Saudino, K. J. (2005). Behavioral genetics and child temperament. *Journal of Developmental & Behavioral Pediatrics, 26*, 214–223.

Sheese, B. E., Voelker, P. M., Rothbart, M. K., & Posner, M. I. (2007). Parenting quality interacts with genetic variation in dopamine receptor DRD4 to influence temperament in early childhood. *Development and Psychopathology, 19*, 1039–1046.

Sheese, B. E., Rothbart, M. K., Posner, M. I., White, L. K., & Fraundorf, S. H. (2008). Executive attention and self-regulation in infancy. *Infant Behavior and Development, 31*, 501–510.

Smolka, M. N., Schumann, G., Wrase, J., Grüsser, S. M., Flor, H., Mann, K., Braus, D. F., Goldman, D., Büchel, C., & Heinz, A. (2005). Catechol-O-methyltransferase val158met genotype affects processing of emotional stimuli in the amygdala and prefrontal cortex. *Journal of Neuroscience, 25*, 836–842.

Spellmann, I., Müller, N., Musil, R., Zill, P., Douhet, A., Dehning, S., Cerovecki, A., Bondy, B., Möller, H. J., & Riedel, M. (2008). Associations of SNAP-25 polymorphisms with cognitive dysfunctions in Caucasian patients with schizophrenia during a brief trail of treatment with atypical antipsychotics. *European Archives of Psychiatry: Clinical Neuroscience, 258*, 335–344.

Stein, M. B., Fallin, M. D., Schork, N. J., & Gelernter, J. (2005). COMT polymorphisms and anxiety-related personality traits. *Neuropsychopharmacology, 30*, 2092–2102.

Tapper, A. R., McKinney, S. L., Nashmi, R., Schwarz, J., Deshpande, P., Labarca, C., Whiteaker, P., Marks, M. J., Collins, A. C., & Lester, H. A. (2004). Nicotine activation of alpha4* receptors: Sufficient for reward, tolerance, and sensitization. *Science, 306*, 1029–1032.

Tsai, S. J., Hong, C. J., Yu, Y. W., & Chen, T. J. (2004). Association study of catechol-O-methyltransferase gene and dopamine D4 receptor gene polymorphisms and personality traits in healthy young Chinese females. *Neuropsychobiology, 50*, 153–156.

Van Tol, H. H., Wu, C. M., Guan, H. C., Ohara, K., Bunzow, J. R., Civelli, O., Kennedy, J., Seeman, P., Niznik, H. B., & Jovanovic, V. (1992). Multiple dopamine D4 receptor variants in the human population. *Nature, 358*, 149–152.

Voelker, P., Sheese, B. E., Rothbart, M. K., & Posner, M. I. (in press). Variations in catechol-o-methyltransferase gene interact with parenting to influence attention in early development. *Neuroscience.*

Van IJzendoorn, M. H., & Bakermans-Kranenburg, M. J. (2006). DRD4 7-repeat polymorphism moderates the association between maternal unresolved loss or trauma and infant disorganization. *Attachment and Human Development, 8*, 291–307.

Winterer, G., Musso, F., Konrad, A., Vucurevic, G., Stoeter, P., Sander, T., & Gallinat, J. (2007). Association of attentional network function with exon 5 variations of the CHRNA4 gene. *Human Molecular Genetics, 16*(18), 2165–2174.

COGNITIVE NEUROPSYCHIATRY
2009, 14 (4/5), 356–376

Psychology Press
Taylor & Francis Group

The genetics of episodic memory

Jeremy Koppel and Terry Goldberg

The Litwin-Zucker Research Center for the Study of Alzheimer's Disease and Memory Disorders, The Feinstein Institute for Medical Research, Manhasset, NY, USA

Introduction: Studies suggest that there is a considerable genetic contribution to individual episodic memory performance. Identifying genes which impact recollection may further elucidate an emerging biology and pave the way towards novel cognitive interventions. To date, several candidate genes have been explored and a few seem to have modest but measurable effects.
Methods: Here we review the biology of memory with particular focus on episodic memory, critically appraise the published evidence supporting the role of several candidate genes, and make suggestions for future pathways of research.
Results: We found moderate evidence for several candidate genes implicated in episodic memory formation, with converging lines of neurobiologic evidence especially strong for only a select few. Perhaps unexpectedly, little work has been done on other aspects of memory, including the semantic and autobiographical systems.
Conclusions: Larger studies utilizing more elaborate methodologies to measure the spectrum of episodic memory are required to move the field forward.

Keywords: Episodic memory; Genetics; Cannabinoids.

INTRODUCTION

In a cogent description William James (1890) considered memory, in a narrow sense with application to its common experience, to be "the knowledge of a former state of mind after it has already once dropped from consciousness; or rather it is the knowledge of an event, or fact, of which we have not been thinking, with the additional consciousness that we have thought or experienced it before" (James, 1890). But more broadly, memory – the entirety of the brain's storage and recollecting function – informs nearly

Correspondence should be addressed to Jeremy Koppel, The Litwin-Zucker Research Center for the Study of Alzheimer's Disease and Memory Disorders, The Feinstein Institute for Medical Research, Manhasset, NY, 11030. E-mail: jkoppel@lij.edu

http://www.psypress.com/cogneuropsychiatry DOI: 10.1080/13546800902990438

every facet of human identity from vocabulary and language; to basic and complex skills and behaviors; to personal experience and the attitudes and opinions that emerge from them. Loss of memory function in any of these capacities – whether through disease or trauma affecting critical brain structures – can have disastrous consequences for afflicted individuals and their families. In this paper, following the outline of some useful concepts in cognitive psychology, we will review the known anatomy and molecular biology of memory, provide a review of the evidence for several of the candidate genes which may be involved in the heritability of memory, and suggest some potential pathways for future research into the genetics of episodic memory.

The literature on memory is rife with a lexicon, at times confusing, distinguishing facets of memory. It reflects a long history of contributions from investigations into cognitive psychology. It is important, before going further, to define terms which delineate phenomena relevant to the discussion of memory biology and genetics. In doing this, we do not mean to suggest that genetics and associated biology (which is the focus of this paper) map uniquely onto the phenomenologic distinctions that are implied with these terms; it is rather a byproduct of the scope of the instruments and methodologies employed in the studies reviewed that require a careful consideration of terminology.

As memory testing is limited by instrumentation, it is imperative to delineate memory into two well established types. In 1972, Endel Tulving proposed a distinction between what he referred to as "episodic" and "semantic" memory (Tulving, 1972). While semantic memory, in this dichotomy, represents knowledge-based memory which has been "over learned", for example who Endel Tulving is or where the Empire State Building is; episodic memory refers to memories derived from experience with a spatiotemporal context, for example what was eaten for breakfast, where it was eaten and how much was paid for it. This is important as testing situations are able to control the conditions of episodic (that depend on experience) but not semantic memory acquisition. Another useful distinction, that has been part of the literature of cognitive psychology dating back to Hermann Ebbinghaus (1850–1909) and William James (1842–1910), describes the spectrum of stability from transience to permanence of memory stores; although many terms have been used to distinguish relative persistence of memory, for the purpose of this discussion we distinguish between (1) immediate memory (past couple seconds to minutes); (2) short-term memory (past couple of minutes to about an hour); and (3) long-term memory (beyond an hour). Obviously testing for short-term or long-term memory involves completely different testing schedules, and as described below, seem to have dichotomous biologies.

THE NEUROBIOLOGY OF MEMORY

Over the past two decades, the neural systems and storage mechanisms that underlie the brain's engrams of experience have been substantially elucidated. The search for candidate genes that impact this type of cognition is both guided by and will further contribute to an emerging understanding of the biology of memory. In reviewing the evidence for particular candidate genes affecting cognitive performance, a sound contextualization of the neurobiology of memory is crucial, as much of that evidence does not directly involve experiments grounded in this biology, but instead relies on human psychometric testing in association with genetic interrogation and is thus left in many cases for biologic interpretation. What follows is a brief recapitulation of the neurobiology of memory, "top down", from anatomy to pathway identification to molecular biology.

The anatomy of memory emerged nearly five decades ago from the clinical recognition of amnestic syndromes which developed following the surgical removal of the medial temporal lobes, most famously in the case of patient H.M. (Gabrieli, Cohen, & Corkin, 1988; Scoville & Milner, 1957). The focus on medial temporal structures became more narrowly focused on hippocampus following the case of patient R.B. who was reported to have memory impairment (reported to be less severe than that experienced by H.M.) following ischemic injury in the absence of other cognitive symptoms, and whose post-mortem examination revealed bilateral lesions of the CA1 field of the hippocampus. (Zola-Morgan, Squire, & Amaral, 1986) Animal model experiments confirmed that bilateral lesions isolated to the hippocampus cause memory impairment, but that as lesions are extended to amygdala and adjacent cortex impairment becomes more severe, and approximates the clinical descriptions of H.M. (Mishkin, 1978; Zola-Morgan & Squire, 1985; Zola-Morgan, Squire, & Amaral, 1986) The amygdala itself seems to be less important in non-affective episodic memory acquisition than surrounding cortical structures (Eichenbaum, Fagan, & Cohen, 1986; Sutherland, McDonald, Hill, & Rudy, 1989). Cortical structures subsequently identified which are critical in memory formation include entorhinal cortex together with perirhinal and parahippocampal cortices (Zola-Morgan, Squire, Amaral, & Suzuki, 1989; Zola-Morgan, Squire, Clower, & Rempel, 1993) As time passes from memory acquisition, evidence suggests that neocortex becomes increasingly important for memory storage and retrieval (Squire & Bayley, 2007; Takashima et al., 2006), and that medial temporal lobe structures play only a "time-limited role" in semantic memory (Smith & Squire, 2009).

Pathways underlying the anatomy within cortical structures implicated in memory acquisition (outlined above) have been mapped. The primary input to the hippocampus is through the entorhinal cortex which acts as a hub for information traveling through the parahippocampal cortex and perirhinal cortex (Squire, 2000). Information then flows through the hippocampus via three principal pathways: the perforant pathway from the entorhinal cortex to the granule cells of the dentate gyrus; the mossy fiber pathway from the granule cells to the pyramidal cells of the CA3 region; and the Schaffer collateral pathway from the CA3 to the CA1 region (Squire, 2000). There is evidence that damage to this hippocampal circuitry may be as disruptive to memory function as more extensive damage to the hippocampus itself (Squire & Zola-Morgan, 1991). A recently published study provides insight into the neuronal correlate of remembrance. Investigators studied free recall in a cohort of patients being treated for intractable epilepsy, and found that at the time of remembrance a subset of neurons in the hippocampus and entorhinal cortex active during memory acquisition of audiovisual sequences exhibited selective firing preceding verbal report of episodic memory (Gelbard-Sagiv, Mukamel, Harel, Malach, & Fried, 2008).

On a more granular level, there is an emerging consensus about the scaffolding on which the molecular biology of memory is built. First described in 1973, long-term potentiation (LTP) is widely viewed as the mechanistic solution to the problem of engram production and the basic building block of memory formation (Bliss & Gardner-Medwin, 1973; Bliss & Lomo, 1973). Within the hippocampus, threshold stimulation of monosynaptic excitatory pathways produce an increase in the efficiency of synaptic transmission; this potentiation can be measured by a persistent increase in the synaptic evoked response of individual cells and even groups of neurons. This enhancement of synaptic transmission relies on the displacement of the Mg^{2+} ion plug from the post-synaptic N-methyl-d-aspartate (NMDA) receptor occupied with glutamate in the context of membrane depolarization (Bliss & Collingridge, 1993). Following this displacement, Ca^{2+} rushes into the postsynaptic cell activating Ca^{2+} calmodulin, several protein kinases, and retrograde signal generators which increase neurotransmitter (glutatmate) release and propagate local LTP. Trafficking of AMPA receptors to the cell membrane also occurs, and promotes early phase LTP. In a method allowing for gene knockout restricted to CA1 neurons in the mouse hippocampus, this model was corroborated by Tsien Huerta, and Tonegawa (1996) who demonstrated in a conditional NMDA receptor knockout mouse that spatial memory formation relies on LTP generated by NMDA-dependent signaling. Thus, the cognitive paradigm of episodic memory emerges from a neurobiology of local situation-dependent signaling events which increase synaptic efficiency.

Consolidation of short-term memories into long-term memories is a separate process and occurs in stages with an attendant dedicated biology; this process has the potential to be interrupted with resultant retrograde amnesia as in the case of traumatic brain injury. The biology of memory consolidation has been substantially elucidated by Kandel and others who have demonstrated that in *Aplysia* converting short-term memories into long-term memories involves inducing phosphorylation via the cAMP-dependent protein kinase (PKA) of transcription factors which modulate the expression of memory genes; the phosphorylated form of CREB-1 induces transcription of memory genes, while CREB-2 is an inhibitory transcription regulator which is thought to block the binding of CREB-1 to its DNA binding site (Abel, Martin, Bartsch, & Kandel, 1998; Dash, Hochner, & Kandel, 1990). It is during the phase when the proteins encoded by CREB-1 activated genes are being generated that consolidation of memory is thought to be most vulnerable to interruption by inhibitors of protein synthesis or head trauma.

The "bioelectric switch" which induces gene-expression and pushes short-term memory into consolidation involves, again, LTP. LTP itself can be divided into an early and a late phase. The early phase which lasts for 1–3 h begins after high-frequency single train cell stimulation in which glutamate binding to the AMPA receptor in the postsynaptic membrane triggers depolarization and unblocking of NMDA receptors with subsequent intracellular calcium influx, a process which does not require protein synthesis. Late phase LTP occurs following several high-frequency action potential trains, requires gene activation, and lasts for over 24 h (Frey, Huang, & Kandel, 1993). In addition to the cAMP→PKA→CREB-1 pathway, late phase LTP is associated with the induction of specific genes some of which, in addition to other functions, potentiate growth of axonal terminals and dendritic spines, facilitating synaptic connections (Baranes et al., 1998; Qian, Gilbert, Colicos, Kandel, & Kuhl, 1993). Studies with mutant mice in which pathways to late but not early phase LTP were interrupted confirm that late phase LTP is crucial only for long-term memory storage, and not for short-term memory (Abel et al., 1997; Bourtchouladze et al., 1998; Pittenger et al., 2002). This suggests that long-term memory is not simply "well encoded" short-term memory, but relies on discrete molecular mechanisms and could potentially have unique expression in cognitive performance variability and pathology. Unfortunately, few human studies have focused on this dichotomy in the biology of mnemonic process.

A REVIEW OF CANDIDATE GENES

Quantifying the magnitude of the influence of genetic variability on cognition in healthy individuals across the life-span is a difficult task, but twin and adoption studies support the notion that cognitive abilities are heritable (Bouchard & McGue, 1981; Plomin, Owen, & McGuffin, 1994). Of note, the heritability of cognitive capacity extends into late-life; in one well designed twin study, modeling estimates suggest that genes account for 62%

of the variance in general cognitive ability and 52% of the variance of domain-specific memory performance in those over 80 years old (McClearn et al., 1997). Identifying the specific genes that form the substrate of this observed influence has been a challenge for the field of cognitive genetics, with consistent evidence for a gene-effect on cognition existing for only a select few genes – among them APOE, BDNF, KIBRA, COMT, 5HT2A – and that with only modest consequence. The paucity of known cognitive genes, taken with the estimated heritability of cognition dictates that there are memory genes yet to be discovered.

The Apo lipoprotein E (APOE) gene codes for a lipid-transporting plasma protein, and has three allelic variations in the human population identified as ε2, ε3, and ε4. The ε4 allele has been consistently found to be a risk-factor for late-onset Alzheimer's disease; meta-analysis of published studies suggests that carriers of ε4 are at a three to four times higher risk of developing AD, while homozygous ε4 carriers have between a ten and twelve times increased risk (Farrer et al., 1997).

There has been enormous interest in this gene as a potential mediator of cognitive performance, particularly episodic memory, in the non-demented population; to some degree, the interest has been driven by the putative biology of non-demented ε4 carriers. Studies have suggested that non-impaired ε4 allele carriers are affected with a diminution of hippocampal volume relative to non-ε4 carriers (Cohen, Small, Lalonde, Friz, & Sunderland, 2001; Moffat, Szekely, Zonderman, Kabani, & Resnick, 2000). There is also evidence from functional neuroimaging that ε4 carriers, prior to the onset of dementia, demonstrate reduced task-associated medial temporal lobe activation (Borghesani et al., 2008) and altered neural activity while performing visual spatial memory tasks (Filbey, Slack, Sunderland, & Cohen, 2006; Mondadori et al., 2007) relative to non-ε4 carriers. Results of investigations into the potential cognitive effect of APOE genotype have been mixed outside the context of AD. While a large number of studies have found an association between the ε4 allele with poor memory performance, with some suggestion of specificity to episodic performance (Bartres-Faz et al., 1999; Bondi et al., 1995; Caselli et al., 1999, 2001; Chey, Kim, & Cho, 2000; Dik et al., 2000; Flory, Manuck, Ferrell, Ryan, & Muldoon, 2000; Helkala et al., 1995; Jonker, Schmand, Lindeboom, Havekes, & Launer, 1998; Mayeux, Small, Tang, Tycko, & Stern, 2001; Nilsson et al., 2006; Staehelin, Perrig-Chiello, Mitrache, Miserez, & Perrig, 1999); there have been studies, including large population-based studies, failing to find such an association (Bennett et al., 2005; Bondi, Salmon, Galasko, Thomas, & Thal, 1999; Bunce, Fratiglioni, Small, Winblad, & Backman, 2004; Jorm et al., 2007; Small et al., 2000; Smith et al., 1998). A meta-analysis of 38 studies published through 2004, including over 20,000 subjects, demonstrated a very modest effect of the ε4 allele on episodic memory (Small, Rosnick, Fratiglioni, & Backman, 2004).

Brain-derived neurotrophic factor (BDNF), a protein found in the central nervous system and the periphery, stimulates the growth and differentiation of neurons and synapses and has been linked to both early and late-phase LTP (Poo, 2001). There is evidence that BDNF is specifically involved in hippocampal LTP (Lu & Gottschalk, 2000), as a consequence of both enhanced synaptic vesicle docking and postsynaptic effects on TrK B receptors (Gottschalk, Pozzo-Miller, Figurov, & Lu, 1998; Jovanovic, Czernik, Fienberg, Greengard, & Sihra, 2000; Levine, Dreyfus, Black, & Plummer, 1995; Pozzo-Miller et al., 1999; Xu et al., 2000). Further, in hippocampal neurons BDNF is produced on demand in an activity dependent manner making it a worthy target of investigation in memory acquisition (Farhadi et al., 2000; Goodman et al., 1996; Mowla et al., 2001). Indeed, the BDNF gene encodes a precursor peptide (proBDNF) which undergoes proteolytic cleavage prior to protein formation (Seidah, Benjannet, Pareek, Chretien, & Murphy, 1996). A common functional single nucleotide polymorphism in the BDNF gene encodes a valine to methionine substitution (Val66Met)in the 5′ pro-BDNF sequence. The biologic and behavioral impact of this SNP on human cognition has been elucidated and provides direct evidence for the importance of BDNF to episodic memory formation. Egan et al. (2003) explored the effects of the Val66Met polymorphism on human episodic memory and expression and secretion of BDNF in cultured hippocampal neurons. In a cohort of 641 subjects, BDNF genotype had a significant effect on episodic, but not semantic or working memory; met/met homozygotes performed significantly worse than heterozygotes or val/val homozygotes in remembering stories. Further, the effects of this polymorphism on intracellular distribution, processing, and secretion of BDNF were examined in hippocampal neurons transfected with either met-BDNF or val-BDNF; while constitutive secretion of BDNF was not altered, met-BDNF neurons exhibited diminished depolarization-dependent secretion (Egan et al., 2003). Additionally, met-BDNF predicted abnormal hippocampal activation on fMRI and lower n-acetyl aspartate on MRI spectroscopy. The insalubrious effect of the met allele on episodic memory has been demonstrated in subsequent studies of the elderly and cohorts with schizophrenia (Dempster et al., 2005; Miyajima et al., 2008). Functionally, an fMRI study revealed that met carriers of the BDNF Val66Met polymorphism have reduced hippocampal engagement during episodic memory processing which, in an interaction with BOLD signal during encoding, was found to account for 25% of the variation in recognition memory performance (Hariri et al., 2003). A subsequent fMRI investigation demonstrated a negative dose-dependent correlation between the met-BDNF allele and localized brain activity (during memory encoding) in the hippocampus bilaterally and the right hippocampal gyrus (Hashimoto et al., 2008). In healthy subjects, the met-BDNF allele was found to confer an 11% reduction in hippocampal volume (Bueller et al., 2006). In a large study of healthy controls and schizophrenic patients met carriers had worse performance on a variety of episodic memory measures and moreover, had reduced gray matter volume in their temporal lobes (Ho et al., 2006).

More recently, investigators used the presumptive neurobiology of BDNF genotypes to make predictions about episodic memory subprocesses (Goldberg et al., 2008). They found that BDNF genotypic effects were equivalent in short and long delay verbal memory tasks, in keeping with the role of BDNF in both early and late-phase LTP. Moreover, based on regional expression patterns and volumetric studies, BDNF genotype has its largest impact on memory acquisition, as opposed to memory monitoring or response bias.

The development and utilization of high-throughput genotyping platforms has led to a paradigm shift in genomics from a candidate gene/hypothesis-based approach to a discovery-based approach; for memory genomics, the result of this approach has been interest in the KIBRA gene. Papassotiropoulos et al. (2006) used Affiymetrix 500K GeneChip SNP genotyping technology to screen pooled DNA from a Swiss cohort of 341 young adults stratified into four subcohorts based on performance on a verbal memory task. The entire cohort was genotyped at 502,627 SNP's. This screening identified a T→C nucleotide substitution within the ninth intron of the KIBRA gene which associated with memory performance. This was validated in the same cohort using a non-pooled, independent genotyping technology (Pyrosequencing). Specifically, carriers of the minor T allele had significantly better free recall performance at 5 min and 24 h after word presentation. The effect of this polymorphism on episodic memory was replicated – as part of the same study – in a second sample of 256 cognitively normal subjects from a cohort in the United States, and in a second Swiss cohort of 424 young adults. To explore the functional relevance of the KIBRA gene, Papassotiropoulos and colleagues used immunocytochemistry and western blotting to detect expression of KIBRA in memory-related human and murine brain structures. The expression levels of the truncated form of KIBRA were high in the human hippocampus; in mice, KIBRA expression was the highest in CA1 region and dentate gyrus of the hippocampus. Finally, to identify KIBRA genotype-dependent differences in brain activation, 15 carriers of the T allele and 15 non-carriers matched for delayed recall performance and Val66Met and 5HT2A genotype (see later) were imaged with fMRI during tasks of episodic memory retrieval. Increased brain activation was observed in non-carriers relative to carriers in the medial temporal lobe, frontal cortex, and parietal cortex, suggesting that non-carriers need increased memory system recruitment to achieve the same level of memory performance. The impact of the KIBRA T→C polymorphism on episodic memory was subsequently replicated in two independent studies of healthy elderly cohorts (Almeida et al., 2008; Schaper, Kolsch, Popp, Wagner, & Jessen, 2008); however, a recent study of two cohorts failed to find any impact of KIBRA on memory performance (Need et al., 2008).

There were several other interesting aspects to this report. Another candidate gene (see below) 5HT2A, was not replicated, raising questions about the utility of genome wide scans versus notions about prior probability as it relates to neurobiological convergent evidence. Second, it was interesting to note that a second SNP CLSTN2, encoding the synaptic protein calsyntenin 2 – highly expressed in brain as opposed to the kidney (i.e., KIBRA) – was found to be highly associated with memory performance in the initial, but not the replication sample, demonstrating the complexities of this type of work and the need for replications.

Dopamine has long been felt to play a critical role in cognition, especially in the prefrontal cortex, an area critical to tasks of executive function such as attention which is a prerequisite for all learning and memory. One enzyme implicated in the catabolism of dopamine is catechol-O-methyltransferase, COMT. A Val158Met substitution in the COMT gene influences the activity and thermal stability of the COMT enzyme, the met allele being less active at physiologic temperatures, which may effect availability of dopamine in the brain (Lotta et al., 1995); the cognitive effects of this polymorphism naturally have been the focus of close attention (Tunbridge, Harrison, & Weinberger, 2006), especially in diseases associated with disturbances in dopamine systems (schizophrenia and Parkinson's disease) and was the subject of a recently published meta-analysis (Barnett, Scoriels, & Munafo, 2008). The effects of the Val158Met substitution on episodic memory in non-disease affected subjects have received less attention. One study found that in healthy middle aged and elderly men, met/met carriers performed better on tests of episodic and semantic memory than val carriers (de Frias et al., 2004). A subsequent study in a cohort of healthy elderly demonstrated that Val158Met heterozygotes performed better on tests of verbal declarative memory than both homozygous groups (Harris et al., 2005). In a separate study of a healthy elderly cohort, there was no association found between COMT genotype and any cognitive measure, including memory (O'Hara et al., 2006). The isolated effects of the COMT genotype on episodic memory in healthy individuals remain unclear, although recent studies suggest that this polymorphism predisposes to risk of substance-induced memory impairment (Schilt et al., 2009), and may play a more fundamental role in prefrontal-hippocampal coupling (Bertolino et al., 2006b). Additionally, from the perspective of biologic phenotypes, there is evidence linking COMT genotype with hippocampal volume (Cerasa et al.,

2008) and blood oxygen-level dependent (BOLD) response elicited by unpleasant stimuli. (Smolka et al., 2005) From an interactive perspective, there is evidence of a transactional/epistatic relationship between the Val158Met genotype and genes encoding important proteins in the dopamine catabolic system, including the dopamine transporter, DAT, on reward behaviors and immediate memory and the activity of the hippocampus during encoding of immediate memory (Bertolino et al., 2006a, 2008; Yacubian et al., 2007).

There is evidence that serotonin systems play a role in learning and memory. Kandel has shown that the neural changes which underlie the sensitization of the gill-withdrawal reflex in Aplysia (presynaptic facilitation) very likely depend on activation of serotonergic neurons (Brunelli, Castellucci, & Kandel, 1976). Loss of 5-HT2A receptors with advancing age has been demonstrated in human brain with positron-emission methodology, specifically effecting hippocampus and medial prefrontal cortex (Sheline, Mintun, Moerlein, & Snyder, 2002). In animals, hippocampus-dependent memory has been shown to be influenced by enhanced tone at serotonin receptors, with enhanced serotonergic transmission abetting both short and long-term memory in rats (Haider, Khaliq, & Haleem, 2007). Further, depletion of tryptophan with resultant diminution of available serotonin impacts memory consolidation (Schmitt et al., 2000). Consonant with this biology, the relationship between a functional polymorphism in the serotonin 5-HT2A receptor and episodic memory has been explored. A common polymorphism in this gene predicts an amino acid substitution (His452Tyr); heterozygotes (His/Tyr) for his polymorphism have demonstrated a less vigorous receptor response to stimulation than major allele homozygotes (His/His) (Ozaki et al., 1997). Three separate association studies have demonstrated that 5-HT2A genotype exerts an influence on episodic memory, with poorer performance in heterozygotes and effect sizes for 30-min free recall reported to be between 0.4 and 0.5 (de Quervain et al., 2003; Reynolds, Jansson, Gatz, & Pedersen, 2006; Wagner, Schuhmacher, Schwab, Zobel, & Maier, 2008). From a structural perspective, in a volumetric MR study, a significant loss of volume in temporal structures was associated with carriers of the minor allele (Filippini et al., 2006).

SUGGESTED PATHWAYS

As reviewed above, several genes have been identified which collectively exert a modest influence on episodic memory; however, if, as suggested by available research, genes account for more than 50% of the variance in episodic memory performance in the population, more memory genes remain to be discovered. In narrowing down the search, the ideal candidate gene(s) would have expression patterns consistent with the anatomy of episodic memory, and some plausible impact on LTP pathways. Using these criteria, available literature suggests a few attractive areas of investigation.

Endogenous cannabinoid receptor genes

One area of interest is the emerging field of endogenous cannabinoid physiology. While the impairing effects of the active component of marijuana, Δ^9–THC, on verbal learning and memory are well established, the relevance of the body's endogenous cannabinoids to learning and memory is not clear (Makela et al., 2006). The lipids anandamide (AEA) and 2-arachidonoyglycerol (2-AG) have been established as ligands in the brain, binding to the cannabinoid (CB_1) receptor; stimulation of this receptor is responsible for the mood altering and cognitive effects of Δ^9–THC (Devane et al., 1992; Sugiura et al., 1995). CB_1 receptor expression patterns are consistent with a role in learning and memory. A high-density of CB_1 receptors are expressed in the hippocampus, with the most significant accumulation in the dentate gyrus, and to a lesser degree the CA3 and CA1 cell layers (Hampson & Deadwyler, 1998; Herkenham, 1991; Herkenham et al., 1991). Given its regional expression profile and potential impact on LTP (reviewed later), the CB_1 receptor gene is a rational target of investigation for an association with episodic memory performance. The CB1 receptor is encoded by CNR1 located on chromosome 6q14-15. A triplet repeat polymorphism in the CNR1 gene has been associated with hebephrenia, a cognitively impaired phenotype of schizophrenia (Ujike et al., 2002). In multiple sclerosis, a disease associated with perturbations in the endocannabinoid system (Di Filippo, Pini, Pelliccioli, Calabresi, & Sarchielli, 2008; Jackson, Diemel, Pryce, & Baker, 2005; Pryce et al., 2003) and cognition (Rao, 1995), CNR1 polymorphisms may impact cognition. One small study reported an association between a synonymous exonic polymorphism in CNR1 and total neurocognitive performance in a test sample of patients with multiple sclerosis, though not in a replication sample (Woolmore et al., 2008).

Surprisingly, given the negative effect of Δ^9–THC on memory, there is evidence that endogenous cannabinoids promote positive changes in neural physiology related to memory, with a potential role in long-term synaptic plasticity; and that, unlike Δ^9–THC, endocannabinoids may have a salutary effect on cognition (Gerdeman & Lovinger, 2003). CB_1 receptors in hippocampus and neocortex are distinctly expressed by GABAergic interneurons, and interact with endocannabinoids produced in post-synaptic neurons in a retrograde manner, with resultant depolarization-induced suppression of inhibition (Wilson & Nicoll, 2002). As modulation of inhibition generally has effects on LTP at excitatory synapses, this physiology argues for cannabinoid mediated enhancement of LTP and, potentially, cognition. This idea was confirmed in an experiment in which the release of endocannabinoids in rat CA1 pyramidal cells were shown to facilitate the induction of LTP in the hippocampus (Carlson, Wang, & Alger, 2002). *In vivo* experiments with mice exploring this hypothetical link between memory and cannabinoid physiology have been equivocal, with reports of both enhanced and impaired memory performance in null mutant CB_1–/– mice (Ledent et al., 1999; Reibaud et al., 1999). However, in a recent stress model of depression in mice, reductions in hippocampal 2-AG levels were associated with deficits in behavioral flexibility (suggested to be related to the inability to forget), implicating hippocampal cannabinoid signaling in the "pruning" of normative memory systems (Hill et al., 2005). Additionally, in a mouse model of Alzheimer's disease, a CB_1 receptor antagonist prevented the amnesia induced by amyloid-beta peptides, suggesting that local production of cannabinoids in Alzheimer's disease contributes to cognitive impairment (Mazzola, Micale, & Drago, 2003). It may be that in lower concentrations endocannabinoids and CB_1 receptor stimulation may promote learning and memory, while this effect may be reversed with increasing cannabinoid tone. More research is required to determine what role the endocannabinoid system and CNR1 gene play in normative cognition.

Calcium signaling genes

Recently, Marambaud and colleagues reported an association between a polymorphisms in a gene preferentially expressed in the hippocampus which encodes a glycoprotein that controls cytosolic calcium concentrations and Alzheimer's disease (Dreses-Werringloer et al., 2008). This gene makes an attractive candidate for future cognitive research for several reasons, and suggests a novel methodology in the search for new genes of cognition. Using TissueInfo (Skrabanek & Campagne, 2001), a high-throughput genomic discovery methodology for tissue expression profiles and specificity, twelve genes were detected with expression restricted to the human hippocampus. Next, these genes were vetted as potential disease candidates using the Alzgene database (Bertram et al., 2000) for localization within known Alzheimer's disease linkage regions, and one gene, subsequently named calcium homeostasis modulator 1 (CALHM1), emerged as a candidate; the confirmed association between CALHM1 and Alzheimer's disease validated this methodology. The CALHM1 P86L substitution associated with Alzheimer's disease was shown in the same report to predict a reduction in calcium ion permeability. As calcium influx in postsynaptic cells drives the cascade that eventuates in synaptic LTP, a gene which controls cytosolic calcium concentrations specifically in the hippocampus might be regarded as a cognition candidate gene until proven otherwise.

Amongst the genes identified in the hippocampus that did not map to Alzheimer's disease linkage regions, several are of interest to cognitive neuroscience. These include CNR1 (reviewed above), Gas1 (growth arrest specific gene 1), and EPHA10 (ephrin A10 receptor). It has been suggested that Gas 1, a gene which has been shown to inhibit cell cycle progression, is induced by activated glutamate receptors and may be involved in cell death related to excitotoxicity (Mellstrom et al., 2002). Altered expression of Gas1 could be hypothesized to affect neuronal integrity in the hippocampus with a downstream impact on cognition. Ephrin receptors, a subfamily of receptor tyrosine kinases, and their ephrin ligands mediate cell–cell communication in neuronal cells directing cell attachment and mobility (Aasheim, Patzke, Hjorthaug, & Finne, 2005). There is emerging evidence implicating ephrin-A receptors, with demonstrated expression in mouse hippocampus, in murine memory models (Gerlai & McNamara, 2000).

CONCLUSIONS

In surveying the existing literature and looking forward, there are several lessons that might inform future research design in cognitive genetics. As the effects in cognitive genetics of individual genes are relatively small, samples need to be large enough to magnify these effects and avoid type II error. Sample sizes less than several hundred for any particular candidate gene are much less likely to find statistically significant differences in cognitive performance. Additionally, replication is required to ensure that very attractive candidates do not suffer from a natural tendency towards type I error born of multiple studies and lack of publication of negative results. Several replications should be regarded as the standard prior to regarding a finding as "real", regardless of the strength of the biology that suggested the investigation. In high-throughput studies, attention needs to be paid to type I error resulting from multiple comparisons without adequate correction. As with any field of research, the cognition study is only as good as its experimental instrument; there should be a focus on developing more sensitive instruments to capture subtle variability in the encoding and retrieval memory pathways that respects memory acquisition as a process with both early and late biologic, and presumably phenomenologic, components. Further work might profitably investigate the cognitive genetics of recollection in memory (versus familiarity), transitivity, and generalization. In tandem with the field of genetics in general, there needs to be a greater appreciation for the increased power and avoidance of confounds that comes with ethnic stratification by ancestry informative markers (Tian, Gregersen, & Seldin, 2008). Studies designed with these guidelines may be more likely to move the field forward.

Classically, the fundamental approach utilized to identify genes of cognition has been to identify candidate genes from illness known to impair cognition, from neurotransmitter systems known to participate in memory-related pathways or from genes fundamental to neuronal integrity followed by hypothesis driven studies to explore their impact on cognition. The candidate gene approach has successfully identified several genes which segregate the performance of healthy individuals, albeit with individually weak allele-specific influences on memory: APOE, BDNF, COMT, and 5HT2A. Recent methodology employs technology driven methodology (high through-put gene sequencing) to identify statistically significant risk-genes, followed by prospective studies to explore impact on cognition. This strategy led to the identification of KIBRA. More focused effort and larger studies are required of both methodologies to move this field forward. Candidate genes aside, the translation of semi-dichotomized memory systems (episodic/semantic or early/late-phase LTP) into published research is largely absent from the field of cognitive genetics. Episodic memory acquisition (as opposed to consolidation), perhaps due to more precise or valid instrumentation, has been the focus of nearly all published memory genetics studies in healthy non-at risk cohorts. Human studies exploiting the recently discovered distinctions between early and late-phase LTP – i.e., gene activation and protein synthesis – have not yet been done. It may be that memory studies with focused methodology that extends over 24 h may have greater sensitivity for variability in performance and less of a "ceiling-effect" while having the advantage of being contemporaneous with the induction of memory-specific genes. Both a candidate gene and discovery-based methodology could be utilized to take advantage of such fundamental neurobiological findings.

REFERENCES

Aasheim, H. C., Patzke, S., Hjorthaug, H. S., & Finne, E. F. (2005). Characterization of a novel Eph receptor tyrosine kinase, EphA10, expressed in testis. *Biochimica et Biophysica Acta, 1723*(1–3), 1–7.

Abel, T., Nguyen, P. V., Barad, M., Deuel, T. A., Kandel, E. R., & Bourtchouladze, R. (1997). Genetic demonstration of a role for PKA in the late phase of LTP and in hippocampus-based long-term memory. *Cell, 88*(5), 615–626.

Abel, T., Martin, K. C., Bartsch, D., & Kandel, E. R. (1998). Memory suppressor genes: Inhibitory constraints on the storage of long-term memory. *Science, 279*(5349), 338–341.

Almeida, O. P., Schwab, S. G., Lautenschlager, N. T., Morar, B., Greenop, K. R., Flicker, L., et al. (2008). KIBRA genetic polymorphism influences episodic memory in later life, but does not increase the risk of mild cognitive impairment. *Journal of Cellular and Molecular Medicine, 12,* 1672–1676.

Baranes, D., Lederfein, D., Huang, Y. Y., Chen, M., Bailey, C. H., & Kandel, E. R. (1998). Tissue plasminogen activator contributes to the late phase of LTP and to synaptic growth in the hippocampal mossy fiber pathway. *Neuron, 21*(4), 813–825.

Barnett, J. H., Scoriels, L., & Munafo, M. R. (2008). Meta-analysis of the cognitive effects of the catechol-O-methyltransferase gene Val158/108Met polymorphism. *Biology and Psychiatry, 64*(2), 137–144.

Bartres-Faz, D., Junque, C., Lopez, A., Valveny, N., Moral, P., Galvez, E., et al. (1999). Apo E influences declarative and procedural learning in age-associated memory impairment. *Neuroreport, 10*(14), 2923–2927.

Bennett, D. A., Schneider, J. A., Wilson, R. S., Bienias, J. L., Berry-Kravis, E., & Arnold, S. E. (2005). Amyloid mediates the association of apolipoprotein E e4 allele to cognitive function in older people. Journal of Neurology. *Neurosurgery and Psychiatry, 76*(9), 1194–1199.

Bertolino, A., Blasi, G., Latorre, V., Rubino, V., Rampino, A., Sinibaldi, L., et al. (2006a). Additive effects of genetic variation in dopamine regulating genes on working memory cortical activity in human brain. *Journal of Neuroscience, 26*(15), 3918–3922.

Bertolino, A., Rubino, V., Sambataro, F., Blasi, G., Latorre, V., Fazio, L., et al. (2006b). Prefrontal-hippocampal coupling during memory processing is modulated by COMT val158met genotype. *Biology and Psychiatry, 60*(11), 1250–1258.

Bertolino, A., Di Giorgio, A., Blasi, G., Sambataro, F., Caforio, G., Sinibaldi, L., et al. (2008). Epistasis between dopamine regulating genes identifies a nonlinear response of the human hippocampus during memory tasks. *Biology and Psychiatry, 64*(3), 226–234.

Bertram, L., Blacker, D., Mullin, K., Keeney, D., Jones, J., Basu, S., et al. (2000). Evidence for genetic linkage of Alzheimer's disease to chromosome 10q. *Science, 290*(5500), 2302–2303.

Bliss, T. V., & Collingridge, G. L. (1993). A synaptic model of memory: Long-term potentiation in the hippocampus. *Nature, 361*(6407), 31–39.

Bliss, T. V., & Gardner-Medwin, A. R. (1973). Long-lasting potentiation of synaptic transmission in the dentate area of the unanaestetized rabbit following stimulation of the perforant path. *Journal of Physiology, 232*(2), 357–374.

Bliss, T. V., & Lomo, T. (1973). Long-lasting potentiation of synaptic transmission in the dentate area of the anaesthetized rabbit following stimulation of the perforant path. *Journal of Physiology, 232*(2), 331–356.

Bondi, M. W., Salmon, D. P., Monsch, A. U., Galasko, D., Butters, N., Klauber, M. R., et al. (1995). Episodic memory changes are associated with the APOE-epsilon 4 allele in nondemented older adults. *Neurology, 45*(12), 2203–2206.

Bondi, M. W., Salmon, D. P., Galasko, D., Thomas, R. G., & Thal, L. J. (1999). Neuropsychological function and apolipoprotein E genotype in the preclinical detection of Alzheimer's disease. *Psychology and Aging, 14*(2), 295–303.

Borghesani, P. R., Johnson, L. C., Shelton, A. L., Peskind, E. R., Aylward, E. H., Schellenberg, G. D., et al. (2008). Altered medial temporal lobe responses during visuospatial encoding in healthy APOE*4 carriers. *Neurobiology of Aging, 29*(7), 981–991.

Bouchard, T. J. Jr., & McGue, M. (1981). Familial studies of intelligence: A review. *Science, 212*(4498), 1055–1059.

Bourtchouladze, R., Abel, T., Berman, N., Gordon, R., Lapidus, K., & Kandel, E. R. (1998). Different training procedures recruit either one or two critical periods for contextual memory consolidation, each of which requires protein synthesis and PKA. *Learning & Memory, 5*(4–5), 365–374.

Brunelli, M., Castellucci, V., & Kandel, E. R. (1976). Synaptic facilitation and behavioral sensitization in Aplysia: Possible role of serotonin and cyclic AMP. *Science, 194*(4270), 1178–1181.

Bueller, J. A., Aftab, M., Sen, S., Gomez-Hassan, D., Burmeister, M., & Zubieta, J. K. (2006). BDNF Val66Met allele is associated with reduced hippocampal volume in healthy subjects. *Biology and Psychiatry, 59*(9), 812–815.

Bunce, D., Fratiglioni, L., Small, B. J., Winblad, B., & Backman, L. (2004). APOE and cognitive decline in preclinical Alzheimer disease and non-demented aging. *Neurology, 63*(5), 816–821.

Carlson, G., Wang, Y., & Alger, B. E. (2002). Endocannabinoids facilitate the induction of LTP in the hippocampus. *Nature Neuroscience, 5*(8), 723–724.

Caselli, R. J., Graff-Radford, N. R., Reiman, E. M., Weaver, A., Osborne, D., Lucas, J., et al. (1999). Preclinical memory decline in cognitively normal apolipoprotein E-epsilon4 homozygotes. *Neurology, 53*(1), 201–207.

Caselli, R. J., Osborne, D., Reiman, E. M., Hentz, J. G., Barbieri, C. J., Saunders, A. M., et al. (2001). Preclinical cognitive decline in late middle-aged asymptomatic apolipoprotein E-e4/4 homozygotes: A replication study. *Journal of Neurology Science, 189*(1–2), 93–98.

Cerasa, A., Gioia, M. C., Labate, A., Liguori, M., Lanza, P., & Quattrone, A. (2008). Impact of catechol-O-methyltransferase Val(108/158) Met genotype on hippocampal and prefrontal gray matter volume. *Neuroreport, 19*(4), 405–408.

Chey, J., Kim, J. W., & Cho, H. Y. (2000). Effects of apolipoprotein E phenotypes on the neuropsychological functions of community-dwelling elderly individuals without dementia. *Neuroscience Letters, 289*(3), 230–234.

Cohen, R. M., Small, C., Lalonde, F., Friz, J., & Sunderland, T. (2001). Effect of apolipoprotein E genotype on hippocampal volume loss in aging healthy women. *Neurology, 57*(12), 2223–2228.

Dash, P. K., Hochner, B., & Kandel, E. R. (1990). Injection of the cAMP-responsive element into the nucleus of Aplysia sensory neurons blocks long-term facilitation. *Nature, 345*(6277), 718–721.

de Frias, C. M., Annerbrink, K., Westberg, L., Eriksson, E., Adolfsson, R., & Nilsson, L. G. (2004). COMT gene polymorphism is associated with declarative memory in adulthood and old age. *Behavioral Genetics, 34*(5), 533–539.

de Quervain, D. J., Henke, K., Aerni, A., Coluccia, D., Wollmer, M. A., Hock, C., et al. (2003). A functional genetic variation of the 5-HT2a receptor affects human memory. *Nature Neuroscience, 6*(11), 1141–1142.

Dempster, E., Toulopoulou, T., McDonald, C., Bramon, E., Walshe, M., Filbey, F., et al. (2005). Association between BDNF val66 met genotype and episodic memory. *American Journal of Medical Genetics Part B: Neuropsychiatric Genetics, 134B*(1), 73–75.

Devane, W. A., Hanus, L., Breuer, A., Pertwee, R. G., Stevenson, L. A., Griffin, G., et al. (1992). Isolation and structure of a brain constituent that binds to the cannabinoid receptor. *Science, 258*(5090), 1946–1949.

Di Filippo, M., Pini, L. A., Pelliccioli, G. P., Calabresi, P., & Sarchielli, P. (2008). Abnormalities in the cerebrospinal fluid levels of endocannabinoids in multiple sclerosis. *Journal of Neurology, Neurosurgery and Psychiatry, 79,* 1224–1229.

Dik, M. G., Jonker, C., Bouter, L. M., Geerlings, M. I., van Kamp, G. J., & Deeg, D. J. (2000). APOE-epsilon4 is associated with memory decline in cognitively impaired elderly. *Neurology, 54*(7), 1492–1497.

Dreses-Werringloer, U., Lambert, J. C., Vingtdeux, V., Zhao, H., Vais, H., Siebert, A., et al. (2008). A polymorphism in CALHM1 influences Ca2+ homeostasis, Abeta levels, and Alzheimer's disease risk. *Cell, 133*(7), 1149–1161.

Egan, M. F., Kojima, M., Callicott, J. H., Goldberg, T. E., Kolachana, B. S., Bertolino, A., et al. (2003). The BDNF val66met polymorphism affects activity-dependent secretion of BDNF and human memory and hippocampal function. *Cell, 112*(2), 257–269.

Eichenbaum, H., Fagan, A., & Cohen, N. J. (1986). Normal olfactory discrimination learning set and facilitation of reversal learning after medial-temporal damage in rats: Implications for an account of preserved learning abilities in amnesia. *Journal of Neuroscience*, *6*(7), 1876–1884.

Farhadi, H. F., Mowla, S. J., Petrecca, K., Morris, S. J., Seidah, N. G., & Murphy, R. A. (2000). Neurotrophin-3 sorts to the constitutive secretory pathway of hippocampal neurons and is diverted to the regulated secretory pathway by coexpression with brain-derived neurotrophic factor. *Journal of Neuroscience*, *20*(11), 4059–4068.

Farrer, L. A., Cupples, L. A., Haines, J. L., Hyman, B., Kukull, W. A., Mayeux, R., et al. (1997). Effects of age, sex, and ethnicity on the association between apolipoprotein E genotype and Alzheimer disease. A meta-analysis. APOE and Alzheimer Disease Meta Analysis Consortium. *Journal of the American Medical Association*, *278*(16), 1349–1356.

Filbey, F. M., Slack, K. J., Sunderland, T. P., & Cohen, R. M. (2006). Functional magnetic resonance imaging and magnetoencephalography differences associated with APOEepsilon4 in young healthy adults. *Neuroreport*, *17*(15), 1585–1590.

Filippini, N., Scassellati, C., Boccardi, M., Pievani, M., Testa, C., Bocchio-Chiavetto, L., et al. (2006). Influence of serotonin receptor 2A His452Tyr polymorphism on brain temporal structures: A volumetric MR study. *European Journal of Human Genetics*, *14*(4), 443–449.

Flory, J. D., Manuck, S. B., Ferrell, R. E., Ryan, C. M., & Muldoon, M. F. (2000). Memory performance and the apolipoprotein E polymorphism in a community sample of middle-aged adults. *American Journal of Medical Genetics*, *96*(6), 707–711.

Frey, U., Huang, Y. Y., & Kandel, E. R. (1993). Effects of cAMP simulate a late stage of LTP in hippocampal CA1 neurons. *Science*, *260*(5114), 1661–1664.

Gabrieli, J. D., Cohen, N. J., & Corkin, S. (1988). The impaired learning of semantic knowledge following bilateral medial temporal-lobe resection. *Brain and Cognition*, *7*(2), 157–177.

Gelbard-Sagiv, H., Mukamel, R., Harel, M., Malach, R., & Fried, I. (2008). Internally generated reactivation of single neurons in human hippocampus during free recall. *Science*, *322*, 96–101.

Gerdeman, G. L., & Lovinger, D. M. (2003). Emerging roles for endocannabinoids in long-term synaptic plasticity. *British Journal of Pharmacoloby*, *140*(5), 781–789.

Gerlai, R., & McNamara, A. (2000). Anesthesia induced retrograde amnesia is ameliorated by ephrinA5-IgG in mice: EphA receptor tyrosine kinases are involved in mammalian memory. *Behavioural in Brain Research*, *108*(2), 133–143.

Goldberg, T. E., Iudicello, J., Russo, C., Elvevag, B., Straub, R., Egan, M. F., et al. (2008). BDNF Val66Met polymorphism significantly affects d' in verbal recognition memory at short and long delays. *Biology in Psychology*, *77*(1), 20–24.

Goodman, L. J., Valverde, J., Lim, F., Geschwind, M. D., Federoff, H. J., Geller, A. I., et al. (1996). Regulated release and polarized localization of brain-derived neurotrophic factor in hippocampal neurons. *Molecular and Cellular Neuroscience*, *7*(3), 222–238.

Gottschalk, W., Pozzo-Miller, L. D., Figurov, A., & Lu, B. (1998). Presynaptic modulation of synaptic transmission and plasticity by brain-derived neurotrophic factor in the developing hippocampus. *Journal of Neuroscience*, *18*(17), 6830–6839.

Haider, S., Khaliq, S., & Haleem, D. J. (2007). Enhanced serotonergic neurotransmission in the hippocampus following tryptophan administration improves learning acquisition and memory consolidation in rats. *Pharmacological Reports*, *59*(1), 53–57.

Hampson, R. E., & Deadwyler, S. A. (1998). Role of cannabinoid receptors in memory storage. *Neurobiology of Disease*, *5*(6 Pt B), 474–482.

Hariri, A. R., Goldberg, T. E., Mattay, V. S., Kolachana, B. S., Callicott, J. H., Egan, M. F., et al. (2003). Brain-derived neurotrophic factor val66met polymorphism affects human memory-related hippocampal activity and predicts memory performance. *Journal of Neuroscience*, *23*(17), 6690–6694.

Harris, S. E., Wright, A. F., Hayward, C., Starr, J. M., Whalley, L. J., & Deary, I. J. (2005). The functional COMT polymorphism, Val 158 Met, is associated with logical memory and the personality trait intellect/imagination in a cohort of healthy 79 year olds. *Neuroscience Letters, 385*(1), 1–6.

Hashimoto, R., Moriguchi, Y., Yamashita, F., Mori, T., Nemoto, K., Okada, T., et al. (2008). Dose-dependent effect of the Val66Met polymorphism of the brain-derived neurotrophic factor gene on memory-related hippocampal activity. *Neuroscience Research, 61*(4), 360–367.

Helkala, E. L., Koivisto, K., Hanninen, T., Vanhanen, M., Kervinen, K., Kuusisto, J., et al. (1995). The association of apolipoprotein E polymorphism with memory: A population based study. *Neuroscience Letters, 191*(3), 141–144.

Herkenham, M. (1991). Characterization and localization of cannabinoid receptors in brain: An in vitro technique using slide-mounted tissue sections. *NIDA – Research Monographs, 112*, 129–145.

Herkenham, M., Lynn, A. B., Johnson, M. R., Melvin, L. S., de Costa, B. R., & Rice, K. C. (1991). Characterization and localization of cannabinoid receptors in rat brain: A quantitative in vitro autoradiographic study. *Journal of Neuroscience, 11*(2), 563–583.

Hill, M. N., Patel, S., Carrier, E. J., Rademacher, D. J., Ormerod, B. K., Hillard, C. J., et al. (2005). Downregulation of endocannabinoid signaling in the hippocampus following chronic unpredictable stress. *Neuropsychopharmacology, 30*(3), 508–515.

Ho, B. C., Milev, P., O'Leary, D. S., Librant, A., Andreasen, N. C., & Wassink, T. H. (2006). Cognitive and magnetic resonance imaging brain morphometric correlates of brain-derived neurotrophic factor Val66Met gene polymorphism in patients with schizophrenia and healthy volunteers. *Archives of General Psychiatry, 63*(7), 731–740.

Jackson, S. J., Diemel, L. T., Pryce, G., & Baker, D. (2005). Cannabinoids and neuroprotection in CNS inflammatory disease. *Journal of Neurology Science, 233*(1–2), 21–25.

James, W. (1890). *Principles of psychology.* New York: Henry Holt and Co.

Jonker, C., Schmand, B., Lindeboom, J., Havekes, L. M., & Launer, L. J. (1998). Association between apolipoprotein E epsilon4 and the rate of cognitive decline in community-dwelling elderly individuals with and without dementia. *Archives of Neurology, 55*(8), 1065–1069.

Jorm, A. F., Mather, K. A., Butterworth, P., Anstey, K. J., Christensen, H., & Easteal, S. (2007). APOE genotype and cognitive functioning in a large age-stratified population sample. *Neuropsychology, 21*(1), 1–8.

Jovanovic, J. N., Czernik, A. J., Fienberg, A. A., Greengard, P., & Sihra, T. S. (2000). Synapsins as mediators of BDNF-enhanced neurotransmitter release. *Nature Neuroscience, 3*(4), 323–329.

Ledent, C., Valverde, O., Cossu, G., Petitet, F., Aubert, J. F., Beslot, F., et al. (1999). Unresponsiveness to cannabinoids and reduced addictive effects of opiates in CB1 receptor knockout mice. *Science, 283*(5400), 401–404.

Levine, E. S., Dreyfus, C. F., Black, I. B., & Plummer, M. R. (1995). Brain-derived neurotrophic factor rapidly enhances synaptic transmission in hippocampal neurons via postsynaptic tyrosine kinase receptors. *Proceedings of the National Academy of Science of the United States of America, 92*(17), 8074–8077.

Lotta, T., Vidgren, J., Tilgmann, C., Ulmanen, I., Melen, K., Julkunen, I., et al. (1995). Kinetics of human soluble and membrane-bound catechol O-methyltransferase: A revised mechanism and description of the thermolabile variant of the enzyme. *Biochemistry, 34*(13), 4202–4210.

Lu, B., & Gottschalk, W. (2000). Modulation of hippocampal synaptic transmission and plasticity by neurotrophins. *Progress in Brain Research, 128*, 231–241.

Makela, P., Wakeley, J., Gijsman, H., Robson, P. J., Bhagwagar, Z., & Rogers, R. D. (2006). Low doses of delta-9 tetrahydrocannabinol (THC) have divergent effects on short-term spatial memory in young, healthy adults. *Neuropsychopharmacology, 31*(2), 462–470.

Mayeux, R., Small, S. A., Tang, M., Tycko, B., & Stern, Y. (2001). Memory performance in healthy elderly without Alzheimer's disease: Effects of time and apolipoprotein-E. *Neurobiology and Aging, 22*(4), 683–689.

Mazzola, C., Micale, V., & Drago, F. (2003). Amnesia induced by beta-amyloid fragments is counteracted by cannabinoid CB1 receptor blockade. *European Journal of Pharmacology, 477*(3), 219–225.

McClearn, G. E., Johansson, B., Berg, S., Pedersen, N. L., Ahern, F., Petrill, S. A., et al. (1997). Substantial genetic influence on cognitive abilities in twins 80 or more years old. *Science, 276*(5318), 1560–1563.

Mellstrom, B., Cena, V., Lamas, M., Perales, C., Gonzalez, C., & Naranjo, J. R. (2002). Gasl is induced during and participates in excitotoxic neuronal death. *Molecular and Cellular Neuroscience, 19*(3), 417–429.

Mishkin, M. (1978). Memory in monkeys severely impaired by combined but not by separate removal of amygdala and hippocampus. *Nature, 273*(5660), 297–298.

Miyajima, F., Ollier, W., Mayes, A., Jackson, A., Thacker, N., Rabbitt, P., et al. (2008). Brain-derived neurotrophic factor polymorphism Val66Met influences cognitive abilities in the elderly. Genes. *Brain and Behavior, 7*(4), 411–417.

Moffat, S. D., Szekely, C. A., Zonderman, A. B., Kabani, N. J., & Resnick, S. M. (2000). Longitudinal change in hippocampal volume as a function of apolipoprotein E genotype. *Neurology, 55*(1), 134–136.

Mondadori, C. R., de Quervain, D. J., Buchmann, A., Mustovic, H., Wollmer, M. A., Schmidt, C. F., et al. (2007). Better memory and neural efficiency in young apolipoprotein E epsilon4 carriers. *Cerebral Cortex, 17*(8), 1934–1947.

Mowla, S. J., Farhadi, H. F., Pareek, S., Atwal, J. K., Morris, S. J., Seidah, N. G., et al. (2001). Biosynthesis and post-translational processing of the precursor to brain-derived neurotrophic factor. *Journal of Biological Chemistry, 276*(16), 12660–12666.

Need, A. C., Attix, D. K., McEvoy, J. M., Cirulli, E. T., Linney, K. N., Wagoner, A. P., et al. (2008). Failure to replicate effect of Kibra on human memory in two large cohorts of European origin. *American Journal of Medical Genetics Part B: Neuropsychiatric Genetics, 147B*(5), 667–668.

Nilsson, L. G., Adolfsson, R., Backman, L., Cruts, M., Nyberg, L., Small, B. J., et al. (2006). The influence of APOE status on episodic and semantic memory: Data from a population-based study. *Neuropsychology, 20*(6), 645–657.

O'Hara, R., Miller, E., Liao, C. P., Way, N., Lin, X., & Hallmayer, J. (2006). COMT genotype, gender and cognition in community-dwelling, older adults. *Neuroscience Letters, 409*(3), 205–209.

Ozaki, N., Manji, H., Lubierman, V., Lu, S. J., Lappalainen, J., Rosenthal, N. E., et al. (1997). A naturally occurring amino acid substitution of the human serotonin 5-HT2A receptor influences amplitude and timing of intracellular calcium mobilization. *Journal of Neurochemistry, 68*(5), 2186–2193.

Papassotiropoulos, A., Stephan, D. A., Huentelman, M. J., Hoerndli, F. J., Craig, D. W., Pearson, J. V., et al. (2006). Common Kibra alleles are associated with human memory performance. *Science, 314*(5798), 475–478.

Pittenger, C., Huang, Y. Y., Paletzki, R. F., Bourtchouladze, R., Scanlin, H., Vronskaya, S., et al. (2002). Reversible inhibition of CREB/ATF transcription factors in region CA1 of the dorsal hippocampus disrupts hippocampus-dependent spatial memory. *Neuron, 34*(3), 447–462.

Plomin, R., Owen, M. J., & McGuffin, P. (1994). The genetic basis of complex human behaviors. *Science, 264*(5166), 1733–1739.

Poo, M. M. (2001). Neurotrophins as synaptic modulators. *Nature Reviews Neuroscience, 2*(1), 24–32.

Pozzo-Miller, L. D., Gottschalk, W., Zhang, L., McDermott, K., Du, J., Gopalakrishnan, R., et al. (1999). Impairments in high-frequency transmission, synaptic vesicle docking, and synaptic protein distribution in the hippocampus of BDNF knockout mice. *Journal of Neuroscience, 19*(12), 4972–4983.

Pryce, G., Ahmed, Z., Hankey, D. J., Jackson, S. J., Croxford, J. L., Pocock, J. M., et al. (2003). Cannabinoids inhibit neurodegeneration in models of multiple sclerosis. *Brain, 126*(Pt 10), 2191–2202.

Qian, Z., Gilbert, M. E., Colicos, M. A., Kandel, E. R., & Kuhl, D. (1993). Tissue-plasminogen activator is induced as an immediate-early gene during seizure, kindling and long-term potentiation. *Nature, 361*(6411), 453–457.

Rao, S. M. (1995). Neuropsychology of multiple sclerosis. *Current Opinion in Neurology, 8*(3), 216–220.

Reibaud, M., Obinu, M. C., Ledent, C., Parmentier, M., Bohme, G. A., & Imperato, A. (1999). Enhancement of memory in cannabinoid CB1 receptor knock-out mice. *European Journal of Pharmacology, 379*(1), R1–2.

Reynolds, C. A., Jansson, M., Gatz, M., & Pedersen, N. L. (2006). Longitudinal change in memory performance associated with HTR2A polymorphism. *Neurobiology and Aging, 27*(1), 150–154.

Schaper, K., Kolsch, H., Popp, J., Wagner, M., & Jessen, F. (2008). KIBRA gene variants are associated with episodic memory in healthy elderly. *Neurobiology and Aging, 29*(7), 1123–1125.

Schilt, T., Koeter, M. W., de Win, M. M., Zinkstok, J. R., van Amelsvoort, T. A., Schmand, B., et al. (2009). The effect of Ecstasy on memory is moderated by a functional polymorphism in the cathechol-O-methyltransferase (COMT) gene. *European Neuropsychopharmacology, 19*(2), 116–124.

Schmitt, J. A., Jorissen, B. L., Sobczak, S., van Boxtel, M. P., Hogervorst, E., Deutz, N. E., et al. (2000). Tryptophan depletion impairs memory consolidation but improves focussed attention in healthy young volunteers. *Journal of Psychopharmacology, 14*(1), 21–29.

Scoville, W. B., & Milner, B. (1957). Loss of recent memory after bilateral hippocampal lesions. Journal of Neurology. *Neurosurgery and Psychiatry, 20*(1), 11–21.

Seidah, N. G., Benjannet, S., Pareek, S., Chretien, M., & Murphy, R. A. (1996). Cellular processing of the neurotrophin precursors of NT3 and BDNF by the mammalian proprotein convertases. *FEBS Letters, 379*(3), 247–250.

Sheline, Y. I., Mintun, M. A., Moerlein, S. M., & Snyder, A. Z. (2002). Greater loss of 5-HT(2A) receptors in midlife than in late life. *American Journal of Psychiatry, 159*(3), 430–435.

Skrabanek, L., & Campagne, F. (2001). TissueInfo: High-throughput identification of tissue expression profiles and specificity. *Nucleic Acids Research, 29*(21), E102–102.

Small, B. J., Graves, A. B., McEvoy, C. L., Crawford, F. C., Mullan, M., & Mortimer, J. A. (2000). Is APOE–epsilon4 a risk factor for cognitive impairment in normal aging? *Neurology, 54*(11), 2082–2088.

Small, B. J., Rosnick, C. B., Fratiglioni, L., & Backman, L. (2004). Apolipoprotein E and cognitive performance: A meta-analysis. *Psychology and Aging, 19*(4), 592–600.

Smith, C. N., & Squire, L. R. (2009). Medial temporal lobe activity during retrieval of semantic memory is related to the age of the memory. *Journal of Neuroscience, 29*(4), 930–938.

Smith, G. E., Bohac, D. L., Waring, S. C., Kokmen, E., Tangalos, E. G., Ivnik, R. J., et al. (1998). Apolipoprotein E genotype influences cognitive "phenotype" in patients with Alzheimer's disease but not in healthy control subjects. *Neurology, 50*(2), 355–362.

Smolka, M. N., Schumann, G., Wrase, J., Grusser, S. M., Flor, H., Mann, K., et al. (2005). Catechol-O-methyltransferase val158met genotype affects processing of emotional stimuli in the amygdala and prefrontal cortex. *Journal of Neuroscience, 25*(4), 836–842.

Squire, L. (2000). *Memory from mind to molecules.* New York: Scientific American Library.

Squire, L. R., & Bayley, P. J. (2007). The neuroscience of remote memory. *Current Opinion in Neurobiology, 17*(2), 185–196.

Squire, L. R., & Zola-Morgan, S. (1991). The medial temporal lobe memory system. *Science, 253*(5026), 1380–1386.

Staehelin, H. B., Perrig-Chiello, P., Mitrache, C., Miserez, A. R., & Perrig, W. J. (1999). Apolipoprotein E genotypes and cognitive functions in healthy elderly persons. *Acta Neurologica Scandinavia*, *100*(1), 53–60.

Sugiura, T., Kondo, S., Sukagawa, A., Nakane, S., Shinoda, A., Itoh, K., et al. (1995). 2-Arachidonoylglycerol: A possible endogenous cannabinoid receptor ligand in brain. *Biochemical and Biophysical Research Communications*, *215*(1), 89–97.

Sutherland, R. J., McDonald, R. J., Hill, C. R., & Rudy, J. W. (1989). Damage to the hippocampal formation in rats selectively impairs the ability to learn cue relationships. *Behavioral and Neural Biology*, *52*(3), 331–356.

Takashima, A., Petersson, K. M., Rutters, F., Tendolkar, I., Jensen, O., Zwarts, M. J., et al. (2006). Declarative memory consolidation in humans: A prospective functional magnetic resonance imaging study. *Proceedings of the National Academy of Science of the United States of America*, *103*(3), 756–761.

Tian, C., Gregersen, P. K., & Seldin, M. F. (2008). Accounting for ancestry: Population substructure and genome-wide association studies. *Human Molecular Genetics*, *17*(R2), R143–150.

Tsien, J. Z., Huerta, P. T., & Tonegawa, S. (1996). The essential role of hippocampal CA1 NMDA receptor-dependent synaptic plasticity in spatial memory. *Cell*, *87*(7), 1327–1338.

Tulving, E. (1972). *Episodic and semantic memory.* New York: Academic Press, Inc.

Tunbridge, E. M., Harrison, P. J., & Weinberger, D. R. (2006). Catechol-o-methyltransferase, cognition, and psychosis: Val158Met and beyond. *Biology and Psychiatry*, *60*(2), 141–151.

Ujike, H., Takaki, M., Nakata, K., Tanaka, Y., Takeda, T., Kodama, M., et al. (2002). CNR1, central cannabinoid receptor gene, associated with susceptibility to hebephrenic schizophrenia. *Molecular Psychiatry*, *7*(5), 515–518.

Wagner, M., Schuhmacher, A., Schwab, S., Zobel, A., & Maier, W. (2008). The His452Tyr variant of the gene encoding the 5-HT2A receptor is specifically associated with consolidation of episodic memory in humans. *International Journal of Neuropsychopharmacology*, *11*, 1163–1167.

Wilson, R. I., & Nicoll, R. A. (2002). Endocannabinoid signaling in the brain. *Science*, *296*(5568), 678–682.

Woolmore, J., Stone, M., Holley, S., Jenkinson, P., Ike, A., Jones, P., et al. (2008). Polymorphisms of the cannabinoid 1 receptor gene and cognitive impairment in multiple sclerosis. *Multiple Sclerosis*, *14*(2), 177–182.

Xu, B., Gottschalk, W., Chow, A., Wilson, R. I., Schnell, E., Zang, K., et al. (2000). The role of brain-derived neurotrophic factor receptors in the mature hippocampus: Modulation of long-term potentiation through a presynaptic mechanism involving TrkB. *Journal of Neuroscience*, *20*(18), 6888–6897.

Yacubian, J., Sommer, T., Schroeder, K., Glascher, J., Kalisch, R., Leuenberger, B., et al. (2007). Gene-gene interaction associated with neural reward sensitivity. *Proceedings of the National Academy of Science of the United States of America*, *104*(19), 8125–8130.

Zola-Morgan, S., & Squire, L. R. (1985). Medial temporal lesions in monkeys impair memory on a variety of tasks sensitive to human amnesia. *Behavioral Neuroscience*, *99*(1), 22–34.

Zola-Morgan, S., Squire, L. R., & Amaral, D. G. (1986). Human amnesia and the medial temporal region: Enduring memory impairment following a bilateral lesion limited to field CA1 of the hippocampus. *Journal of Neuroscience*, *6*(10), 2950–2967.

Zola-Morgan, S., Squire, L. R., Amaral, D. G., & Suzuki, W. A. (1989). Lesions of perirhinal and parahippocampal cortex that spare the amygdala and hippocampal formation produce severe memory impairment. *Journal of Neuroscience*, *9*(12), 4355–4370.

Zola-Morgan, S., Squire, L. R., Clower, R. P., & Rempel, N. L. (1993). Damage to the perirhinal cortex exacerbates memory impairment following lesions to the hippocampal formation. *Journal of Neuroscience*, *13*(1), 251–265.

COGNITIVE NEUROPSYCHIATRY
2009, 14 (4/5), 377–390

Psychology Press
Taylor & Francis Group

Epigenetics, genomic mutations and cognitive function

Abraham Reichenberg[1], Jonathan Mill[2], and James H. MacCabe[1]

[1]*Division of Psychological Medicine and Psychiatry, Institute of Psychiatry, King's College London, London, UK, and* [2]*MRC Social Genetic and Developmental Psychiatry Center, Institute of Psychiatry, King's College London, London, UK*

Introduction. There is growing interest in the role of single genes in cognitive functions. Association studies are the most commonly applied method in this field. This method assumes that the genetic information affecting cognitive processes is "static" and unchanging. However, there is accumulating evidence that dynamic genomic and epigenetic alterations can modulate complex cognitive processes, and influence susceptibility to disorders associated with impaired cognitive functioning.
Methods. We present an overview of genomic and epigenetic mechanisms, and discuss the cognitive and psychiatric consequences of genomic and genetic abnormalities.
Results. Genomic and epigenetic changes can affect complex cognitive functions, including learning and memory and are causative in several developmental and psychiatric disorders effecting language, social functioning and IQ.
Conclusions. Genomic and epigenetic disorders are "experiments of nature" that offer unique and valuable insight in to the physiology of general and specific cognitive functions.

Keywords: Cognition; Epigenetics; Genomics; Mutations; CNV.

INTRODUCTION

The most commonly applied method to evaluate the potential role of individual genes in cognitive processes is using association studies, in which the strength of the relationship between the variants of a specific candidate gene and phenotype (cognition) is tested (Goldberg & Weinberger, 2004). This method assumes that the genetic information affecting cognitive processes is "static" and unchanging.

Correspondence should be addressed to Abraham (Avi) Reichenberg, Division of Psychological Medicine and Psychiatry, Institute of Psychiatry, King's College London, Room M7.02.04A, De Crespigny Park, London SE5 8AF, UK. E-mail: avi.reichenberg@iop.kcl.ac.uk

http://www.psypress.com/cogneuropsychiatry DOI: 10.1080/13546800902978417

However, numerous epidemiological, clinical, and molecular features associated with complex cognitive processes are difficult to explain using this traditional gene-based model. These include the incomplete concordance between MZ twins in cognitive performance, gender differences in cognitive abilities, a fluctuating course of cognitive disorders with periods of remission and relapse, peaks of susceptibility to disease coinciding with hormonal changes, and parent-of-origin effects (Mill & Petronis, 2007; Mill et al., 2008). These observations have led to speculation about the importance of dynamic genomic and epigenetic factors in modulating complex cognitive processes such as learning and the formation of memories, and in influencing susceptibility to disorders associated with impaired cognitive functioning.

The following review presents a brief background to the dynamic genomic and epigenetic modifications occurring in the human genome, and the causes and consequences of several disorders of genomic and epigenetic origin that result in a disruption in human cognitive abilities. Although our understanding of these processes is incomplete and is still under extensive investigation, we discuss implications for research on cognition and psychiatric disorders.

GENOMIC MODIFICATION

Several recent studies have highlighted the high number of copy number variations (CNVs), comprising deletions, insertions, duplications, repeats and other complex multi-site variants, that occur in the genome (Redon et al., 2006) (see Figure 1). CNVs are defined as DNA segments of 1 kb or larger which are present in variable copy numbers in comparison with a reference genome (Feuk, Carson, & Scherer, 2006; Feuk, Marshall, Wintle, & Scherer, 2006). CNV polymorphisms are more dynamic than point mutations to the DNA sequence, and often occur *de novo.* Whilst it has been known for some time that gene copy number is often elevated in various types of cancer

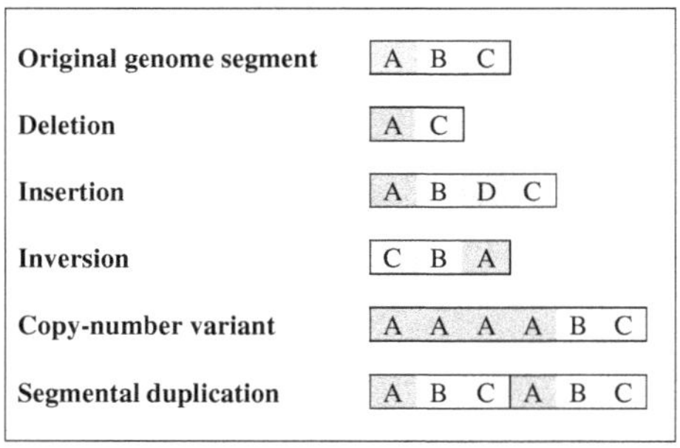

Figure 1. Structural variations in the human genome.

and that chromosomal rearrangements predispose to mental retardation (Knight et al., 1999), copy number variation has also been recently associated with various complex neuropsychiatric cognitive disorders such as autism and schizophrenia (Cook & Scherer, 2008; Sebat et al., 2007) (see later).

One type of CNVs, trinucleotide repeats, have been linked to a number of genetic disorders that do not appear to follow classical Mendelian inheritance patterns such as Fragile X syndrome (see below). These trinucleotide repeat disorders are caused by problems in recombination and replication during meiosis. In most trinucleotide repeat disorders, the larger the expansion the more pathogenic the effects of the repeat; the severity of disease generally increases with length. For most trinucleotide repeat disorders, there is a "threshold" repeat length, above which the phenotypic effects of the expansion become more severe.

EPIGENETIC MODIFICATION

Epigenetics refers to the reversible regulation of various genomic functions, occurring independently of DNA sequence, mediated principally through changes in DNA methylation and chromatin structure (Jaenisch & Bird, 2003). Epigenetic processes are essential for normal cellular development and differentiation, and allow the long-term regulation of gene function through non-mutagenic mechanisms (Henikoff & Matzke, 1997).

Like the DNA sequence, the epigenetic profile of somatic cells is inherited during mitosis. Unlike the DNA sequence, which is stable and strongly conserved, epigenetic processes can be highly dynamic: they are tissue-specific, developmentally regulated, and often induced by exposure to a range of external environmental factors (Dolinoy, Weidman, & Jirtle, 2007).

Cytosine methylation is the best understood and most stable epigenetic modification modulating the transcriptional plasticity of mammalian genomes (Figure 2). It is intrinsically linked to the regulation of gene expression, with many genes demonstrating an inverse correlation between the degree of promoter DNA methylation and the level of expression (Jaenisch & Bird, 2003). The methylation of CpG sites in the promoter regulatory regions of many genes disrupts the binding of transcription factors and attracts methyl-binding proteins that initiate chromatin compaction and gene silencing.

The post-translational modification of histones, the basic proteins around which DNA is wrapped to form nucleosomes, comprises the other major type of epigenetic mechanism related to gene expression (Figure 2). A number of covalent histone modifications, occurring at specific residues, have been described (e.g., acetylation, methylation, phosphorylation, SUMOylation, and ubiquitylation), which together constitute a complex "histone code"

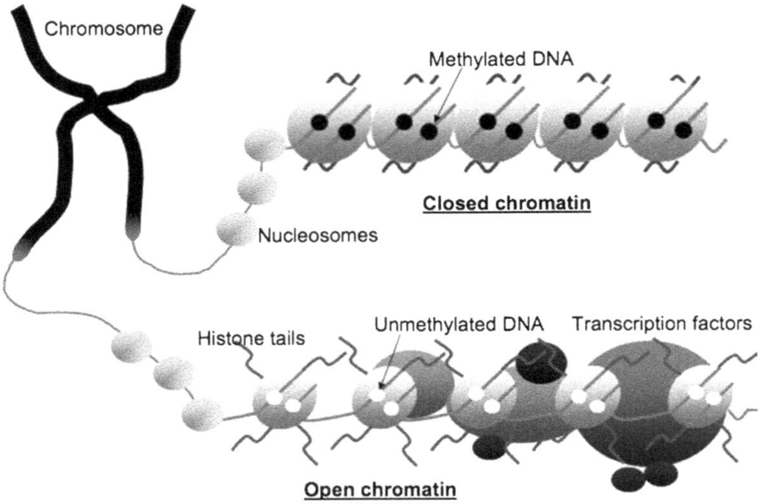

Figure 2. The main components of the epigenetic code. The two main components of the epigenetic code are DNA methylation and Histone modification. DNA Methyl marks added to certain DNA bases repress gene activity. A combination of different molecules can attach to the "tails" of proteins called histones. These alter the activity of the DNA wrapped around them.

modulating gene expression via alterations in chromatin structure (Berger, 2007).

While often investigated independently, epigenetic modifications to DNA and histones are not mutually exclusive, and clearly interact in a number of ways; it is becoming apparent that the classification of epigenetic mechanisms in terms of either gene activation or suppression might be too simplistic (Berger, 2007).

Recently, a third epigenetic system involving small interfering RNA (siRNA) has been described (Hamilton, Voinnet, Chappell, & Baulcombe, 2002). It has been shown that siRNA can suppress the activity of specific genes via targeted RNA interference (RNAi), a mechanism important in the developmental regulation of gene expression. RNAi is known to cause epigenetic changes in gene transcription, mediated by both DNA methylation and histone modifications.

COGNITIVE PHENOTYPE OF GENOMIC AND EPIGENETIC DISORDERS

Several developmental conditions, including common forms of mental retardation, can be attributed, at least in part, to dynamic disruption in the brain's genomic and epigenetic function.

COGNITIVE PHENOTYPE IN GENOMIC DISORDERS

The term "genomic disorder" is typically used to describe a gain (duplication) or loss (deletion) of a specific chromosomal region, associated with a clinical genetic syndrome that may present with congenital anomalies, or with impairment in neurological and cognitive function. In general, the clinical features are believed to reflect changes in normal copy number or dosage of the genes contained within a given genomic interval, one or more of which contribute to the resulting phenotype (Emanuel, 2008).

Williams syndrome

Williams syndrome occurs in 1 of every 20,000 live births (Stromme, Bjornstad, & Ramstad, 2002), and arises from deletions in chromosome 7 (7q11.23) that typically include multiple genes (approximately 28 genes).

Williams syndrome patients exhibit an interesting idiosyncratic social behavioral syndrome – high sociability and empathy for others. Typically, Williams syndrome patients are socially fearless, engaging eagerly in social interaction even with strangers. Intriguingly, this remarkable hypersociability is coupled with a strong undercurrent of anxiety that relates to non-social objects (Meyer-Lindenberg, Mervis, & Berman, 2006).

Williams syndrome is associated with mild to moderate mental retardation. In particular, a severe visuospatial construction deficit is a fundamental stable phenotype in Williams syndrome, contrasting with a relative strength in verbal short-term memory and language. Attention deficit hyperactivity disorder (ADHD) is also common (Meyer-Lindenberg, Mervis, & Berman, 2006).

Because of the large number of genes effected by the full deletion of the chromosome, patients can be found with only "partial" deletions and this can be used for clarifying more specific genotype/phenotype relationships (e.g., Meyer-Lindenberg, Mervis, & Berman, 2006). Williams syndrome thus provides an interesting example of genetic contributions to complex social behaviors (Deutsch, Rosse, & Schwartz, 2007).

22q11 Deletion syndrome (Velocardiofacial/DiGeorge syndrome)

A deletion at chromosome 22q11 is the most frequently known deletion found in humans, occurring in approximately one of every 4000 live births. Its occurrence is associated with a characteristic facial dysmorphology, a range of congenital abnormalities, and cognitive psychiatric problems, especially psychosis (Williams & Owen, 2004). The 22q11 deletion syndrome is associated with a high frequency of learning disabilities. Patients have relatively preserved verbal IQ in comparison to performance IQ. The relative strength in verbal abilities is evident despite speech and language impairments which are common in the disorder (Wang et al., 1998). More specific cognitive impairments have been characterized, especially deficits in visual-spatial memory (Bearden et al., 2001).

22q13 Deletion syndrome

22q13.3 deletions result from the loss of genetic material from the terminus of the long arm of one copy of chromosome 22. Monosomy 22q13.3 can accompany a simple deletion, an unbalanced translocation, or a ring chromosome. About 75% of all cases are simple deletions, with the remaining 25% of cases resulting from structural rearrangements of the affected segment. Deletion 22q13.3 is usually a de novo finding; however, approximately 20% of cases are familial, in which a child receives an unbalanced chromosome complement from a parent who carries a balanced reciprocal translocation (Cusmano-Ozog, Manning, & Hoyme, 2007). Bonaglia et al. (2001) suggested that proline rich synapse associated protein 2 (SHANK3) was a good candidate gene for the 22q13.3 deletion syndrome, as it is preferentially expressed in the cerebral cortex and cerebellum, and encodes a scaffolding protein involved in the postsynaptic density of excitatory synapses.

Mental retardation is frequently observed, with the majority of individuals in the moderate to severe range. Speech delay or severe language impairment with little to no speech is common. In addition to mental retardation and speech delay, most patients meet diagnostic criteria for autism and demonstrate behavioral deficits including: poor eye contact, stereotypical behaviors, self-stimulation, repetitive chewing behaviors, bruxism, biting, hitting and abnormal sleep patterns translocation (Cusmano-Ozog, Manning, & Hoyme, 2007).

Deletion on chromosome 1q21.1

De novo and inherited deletions on chromosome 1q21.1 have recently been associated with a broad range of developmental and psychiatric disorders, including mental retardation, autism (Mefford et al., 2008) and schizophrenia (International Schizophrenia Consortium, 2008; Stefansson et al., 2008). It has been suggested that deletions on chromosome 1q21.1 are evident in 0.5% of persons with developmental abnormalities, and 0.26% of patients with schizophrenia. Importantly, this deletion is characterized by a high degree of phenotypic variability and lack of distinct syndromic features demonstrating some of the challenges in understanding the role of structural variants in human disease and cognition (Mefford et al., 2008). Investigation of genetic and environmental modifiers is therefore necessary to fully understand etiological processes of genomic disorders.

COGNITIVE PHENOTYPE IN EPIGENETIC DISORDERS

Mutations in genes that affect global epigenetic profiles can give rise to human diseases, which can be inherited or somatically acquired (Egger, Liang, Aparicio, & Jones, 2004). Many of these epigenetic abnormalities result in cognitive disabilities. Interestingly, many of these "epigenetic" disorders also involve structural changes to the genome, for example the expanded triplet repeats of Fragile X and the deletion on chromiosone 15q in Angelman syndrome. This suggests a common mechanism by which the gross physical changes causing genomic disorders may exert their effects.

Fragile X syndrome

The fragile X syndrome is caused by an expanded triplet repeat (CGG) on the X chromosome (Xq27.3). The disorder is called fragile X because the many repeats cause the chromosome to be fragile at that point and to break during laboratory preparation of chromosomes. The frequency of fragile X is estimated as 1 in 5000 males and 1 in 10,000 females (Crawford, Acuna, & Sherman, 2001).

The triplet repeat is in an untranslated region at the beginning of a gene (Fragile X Mental Retardation-1, FMR1) that, when expanded to a full mutation, prevents that gene from being transcribed. The full mutation for fragile X, with its hundreds of CGG repeats, causes hypermethylation and thus shuts down transcription of the FMR1 gene (Plomin, DeFries, McClearn, & McGuffin, 2008). This leads to the absence of Fragile X Mental Retardation Protein (FMRP), a molecule that is widely expressed in brain and has been shown to regulate translation of mRNAs important for synaptic plasticity and neuronal maturation (Penagarikano, Mulle, & Warren, 2007).

Most cases of fragile X males are moderately disabled, but many are only mildly disabled and some have normal intelligence. Only about one-half of girls with fragile X are affected, because one of the two X chromosomes for girls could be normal. In addition to mental retardation, learning disabilities, hyperactivity, aggressiveness, and autism-like symptoms are also common (Spreen, Risser, & Edgell, 1995). For fragile X males, IQ declines during adolescence have been reported (Spreen, Risser, & Edgell, 1995).

The study of the molecular basis of Fragile X syndrome has translated to some promising therapeutic strategies to reactivate the silenced FMR-1 gene, or to compensate for the absence of FMRP. The latter approach has proved effective in reducing deficits in several animal models of Fragile X, and clinical trials are planned in humans (Penagarikano et al., 2007).

Rett syndrome

Rett syndrome is the most common single-gene cause of general cognitive disability in females (1 in 10,000). The disorder shows few effects in infancy, although the head, hands, and feet are slow to grow. Cognitive development is normal during infancy but, by school age, girls with Rett syndrome are generally unable to talk and about half are unable to walk (Weaving, Ellaway, Gecz, & Christodoulou, 2005).

The disorder was mapped to the long arm of the X chromosome (Xq28), and then to a specific gene (MECP2, which encodes methyl-CpG-binding protein-2) (Amir et al., 1999). MECP2 is a gene involved in the methylation process that silences other genes during development and thus has diffuse effects throughout the brain (Bienvenu & Chelly, 2006). Males with MECP2 mutations usually die before or shortly after birth (Plomin, DeFries, McClearn, & McGuffin, 2008).

The disorder is characterized by severe to profound mental retardation, stereotyped movements and diminished social interests, as well as impairments in expressive and receptive language (American Psychiatric Association, 1994).

Angelman syndrome

Angelman syndrome is a severe form of mental retardation that occurs in 1 of every 15,000–20,000 births. It results from deregulation of one or more imprinted genes at 15q11–13 (Weeber, Levenson, & Sweatt, 2002).

It is often accompanied by a happy disposition with bouts of inappropriate laughter, hyperactivity, and sleep disorders (Laan, v Haeringen, & Brouwer, 1999). All patients have severe mental retardation and delayed motor milestones. Language does not develop, with most patients having a vocabulary of only one or two words despite having reasonable comprehension of simple commands and sentences (Laan, v Haeringen, & Brouwer, 1999).

Prader-Willi syndrome

The incidence of Prader-Willi syndrome is estimated to be between 1:10,000 and 1:25,000 births. It is caused by a faulty genomic imprinting on chromosome 15q11– defined as parent-specific, monoallelic expression of a gene (Egger, Liang, Aparicio, & Jones, 2004). In such conditions, an abnormal phenotype is established as a result of the absence of the paternal or maternal copy of an imprinted gene or because of deregulation of an imprinted gene. The majority of cases (about 70%) are caused by a *de novo* deletion on chromosome 15 inherited from the father, whilst about 25–30% are caused by inheriting two chromosome 15 s from the mother, instead of one from the mother and one from the father (maternal disomy).

Most people with Prader-Willi syndrome have borderline or moderate learning difficulties. The average IQ is around 70, but some have normal IQ. Language impairments are evident, as well as impairments in memory, attention and reasoning. Visual organization and perception are good (Holm et al., 1993).

EPIGENETICS IN LEARNING AND MEMORY PROCESSES

The formation of long-term memories is thought to involve rapid changes in gene expression, and there is growing evidence that histone modifications and DNA methylation may be involved (Tsankova, Renthal, Kumar, & Nestler, 2007). Animal studies have documented a role for histone modifications processes in hippocampus-dependent memory formation, including spatial mazes, contextual fear conditioning and novel object recognition (Levenson & Sweatt, 2005). Synaptic plasticity is believed to contribute to the formation of long-term memories and also seems to have an epigenetic component (Guan et al., 2002).

Recent findings have also implicated changes in DNA methylation in learning and memory. In contrast to the more traditional view that DNA methylation is a highly stable modification, they suggest that DNA methylation may be subject to rapid and dynamic regulation in the nervous system (Tsankova, Renthal, Kumar, & Nestler, 2007). Contextual fear conditioning induced DNA methyltransferase expression has been documented in the hippocampus. Fear conditioning also induced demethylation of the reelin promoter, indicating that both DNA methylation and demethylation are highly regulated (for review see Tsankova, Renthal, Kumar, & Nestler, 2007).

GENOMIC AND EPIGENETIC CHANGES IN COMPLEX POLYGENETIC COGNITIVE-BEHAVIORAL DISORDERS

Psychiatric disorders have a complex phentoype of cognitive and behavioral impairments, resulting from a multifactorial polygenetic and environmental etiology. Schizophrenia, in particular, is characterized by severe cognitive and social impairment (Reichenberg & Harvey, 2007; Reichenberg et al., 2002; MacCabe et al., 2008). Genomic and epigenetic research in psychiatry is still in its infancy, and relatively few studies have empirically investigated the role of such factors in schizophrenia.

Four very recent studies, all using different designs, have provided evidence that schizophrenia is associated with de novo copy number variations (International Schizophrenia Consortium, 2008; Stefansson et al., 2008; Walsh et al., 2008; Xu et al., 2008). For example, Walsh et al. (2008) performed genome-wide scans in 418 individuals with schizophrenia and 268 healthy controls, for CNVs of > 100 kb. They identified 53 such CNVs that had not been previously reported, and these were three times more common in cases of schizophrenia than controls. Furthermore, the novel CNVs detected in the cases, but not those in controls, were overrepresented in pathways important in neurodevelopment. These data imply that rare *de novo* mutations are responsible for a higher proportion of the genetic risk for schizophrenia than previously assumed, and this may help to explain why schizophrenia persists in the population despite severely impairing fertility (MacCabe et al., 2009).

The possible role of such mutations in cognition and social development is still to be studied in detail, but examples for this approach already exists: The 22q11 deletion syndrome is characterized by adult-onset psychosis in around 30% of cases (Murphy, Jones, & Owen, 1999), coupled with intellectual deterioration (Gothelf et al., 2005). Of particular interest, the catechol-O-methyltransferase (COMT) gene is located in the deleted region. COMT is responsible for metabolising catecholamines, including dopamine, and been studied extensively as a candidate gene for schizophrenia (Williams, Owen, & O'Donovan, 2007), and for executive control functions (Egan et al., 2001).

A recent study of post-mortem brain tissue from patients with bipolar disorder and schizophrenia uncovered evidence for DNA methylation differences in numerous loci, including several genes that have been functionally linked to disease etiology (Mill et al., 2008). Consistent with increasing evidence for altered glutamatergic and GABAergic neurotransmission in the pathogenesis of major psychosis, epigenetic changes were identified in loci associated with both these neurotransmitter pathways. Glutamate and GABA have important roles in higher-order cognitive and social functioning (Gray & Roth, 2007; Tan et al., 2007).

SUMMARY AND CONCLUSIONS

Genomic and epigenetic research in psychiatry and cognitive psychology is still in its infancy. Genetic and epigenetic disorders are "experiments of nature" that offer unique and valuable insight in to the physiology of general and specific cognitive functions. The various genetic and epigenetic syndromes present a unique opportunity to study neuropsychology, cognitive development and neurobiology. In particular, general intellectual ability, language and social cognition.

Major advances in the understanding of cognitive development could come from delineating the combined contribution of the different genes affected in genomic and epigenetic disorders. Several complementary strategies can be followed. The detection and study of individuals with atypical deletions (e.g., partial deletions in Williams syndrome) – which, although rare, sometimes occur as a result of the complexity of genetic structures – may allow making inferences about contributions of groups of genes. Mouse knockout models can be developed to allow the investigation of single gene, and gene interaction effects in animals. With the development of novel microarray-based "epigenomic" techniques, it is now feasible to investigate epigenetic and genomic modifications on a global scale across the genome, including epigenetic and genomic cognitive functions association studies.

REFERENCES

American Psychiatric Association (1994). *Diagnostic and statistical manual of mental disorders* (4th ed.). Washington, DC: APA.

Amir, R. E., Van den Veyver, I. B., Wan, M., Tran, C. Q., Francke, U., & Zoghbi, H. Y. (1999). Rett syndrome is caused by mutations in X-linked MECP2, encoding methyl-CpG-binding protein 2. *Nature Genetitcs, 23*(2), 185–188.

Bearden, C. E., Woodin, M. F., Wang, P. P., Moss, E., McDonald-McGinn, D., Zackai, E., et al. (2001). The neurocognitive phenotype of the 22q11.2 deletion syndrome: Selective deficit in visual-spatial memory. *Journal of Clinical and Experimental Neuropsychology, 23*(4), 447–464.

Berger, S. L. (2007). The complex language of chromatin regulation during transcription. *Nature, 447*(7143), 407–412.

Bienvenu, T., & Chelly, J. (2006). Molecular genetics of Rett syndrome: When DNA methylation goes unrecognized. *Nature Reviews Genetics, 7*(6), 415–426.

Bonaglia, M. C., Giorda, R., Borgatti, R., Felisari, G., Gagliardi, C., Selicorni, A., et al. (2001). Disruption of the ProSAP2 gene in a t(12;22)(q24.1;q13.3) is associated with the 22q13.3 deletion syndrome. *American Journal of Human Genetics, 69*(2), 261–268.

Consortium, I. S. (2008). Rare chromosomal deletions and duplications increase risk of schizophrenia. *Nature, 455*(7210), 237–241.

Cook Jr, E. H., & Scherer, S. W. (2008). Copy-number variations associated with neuropsychiatric conditions. *Nature, 455*(7215), 919–923.

Crawford, D. C., Acuna, J. M., & Sherman, S. L. (2001). FMR1 and the fragile X syndrome: Human genome epidemiology review. *Genetics in Medicine, 3*(5), 359–371.

Cusmano-Ozog, K., Manning, M. A., & Hoyme, H. E. (2007). 22q13.3 deletion syndrome: A recognizable malformation syndrome associated with marked speech and language delay. *American Journal of Medical Genetics Part C: Seminars in Medical Genetics, 145C*(4), 393–398.

Deutsch, S. I., Rosse, R. B., & Schwartz, B. L. (2007). Williams syndrome: A genetic deletion disorder presenting clues to the biology of sociability and clinical challenges of hypersociability. *CNS Spectrums*, *12*(12), 903–907.

Dolinoy, D. C., Weidman, J. R., & Jirtle, R. L. (2007). Epigenetic gene regulation: Linking early developmental environment to adult disease. *Reproductive Toxicology*, *23*(3), 297–307.

Durand, C. M., Betancur, C., Boeckers, T. M., Bockmann, J., Chaste, P., Fauchereau, F., et al. (2007). Mutations in the gene encoding the synaptic scaffolding protein SHANK3 are associated with autism spectrum disorders. *Nature Genetics*, *39*(1), 25–27.

Egan, M. F., Goldberg, T. E., Kolachana, B. S., Callicott, J. H., Mazzanti, C. M., Straub, R. E., et al. (2001). Effect of COMT Val108/158 Met genotype on frontal lobe function and risk for schizophrenia. *Proceedings of the National Academy of Sciences of the United States of America*, *98*(12), 6917–6922.

Egger, G., Liang, G., Aparicio, A., & Jones, P. A. (2004). Epigenetics in human disease and prospects for epigenetic therapy. *Nature*, *429*(6990), 457–463.

Emanuel, B. S. (2008). Molecular mechanisms and diagnosis of chromosome 22q11.2 rearrangements. *Developmental Disabilities Research Reviews*, *14*(1), 11–18.

Feuk, L., Carson, A. R., & Scherer, S. W. (2006). Structural variation in the human genome. *Nature Reviews Genetics*, *7*(2), 85–97.

Feuk, L., Marshall, C. R., Wintle, R. F., & Scherer, S. W. (2006). Structural variants: Changing the landscape of chromosomes and design of disease studies. *Human Molecular Genetics*, *15* (Review Issue No. 1), R57–66.

Goldberg, T. E., & Weinberger, D. R. (2004). Genes and the parsing of cognitive processes. *Trends in Cognitive Science*, *8*(7), 325–335.

Gothelf, D., Eliez, S., Thompson, T., Hinard, C., Penniman, L., Feinstein, C., et al. (2005). COMT genotype predicts longitudinal cognitive decline and psychosis in 22q11.2 deletion syndrome. *Nature Neuroscience*, *8*(11), 1500–1502.

Gray, J. A., & Roth, B. L. (2007). Molecular targets for treating cognitive dysfunction in schizophrenia. *Schizophrenia Bulletin*, *33*(5), 1100–1119.

Guan, Z., Giustetto, M., Lomvardas, S., Kim, J. H., Miniaci, M. C., Schwartz, J. H., et al. (2002). Integration of long-term-memory-related synaptic plasticity involves bidirectional regulation of gene expression and chromatin structure. *Cell*, *111*(4), 483–493.

Hamilton, A., Voinnet, O., Chappell, L., & Baulcombe, D. (2002). Two classes of short interfering RNA in RNA silencing. *EMBO Journal*, *21*(17), 4671–4679.

Henikoff, S., & Matzke, M. A. (1997). Exploring and explaining epigenetic effects. *Trends in Genetics*, *13*(8), 293–295.

Holm, V. A., Cassidy, S. B., Butler, M. G., Hanchett, J. M., Greenswag, L. R., & Whitman, B. Y. et al. (1993). Prader-Willi syndrome: Consensus diagnostic criteria. *Pediatrics*, *91*(2), 398–402.

International Schizophrenia Consortium (2008). Rare chromosomal deletions and duplications increase risk of schizophrenia. *Nature*, *455*(7210), 237–241.

Knight, S. J., Regan, R., Nicod, A., Horsley, S. W., Kearney, L., Homfray, T., et al. (1999). Subtle chromosomal rearrangements in children with unexplained mental retardation. *Lancet*, *354*(9191), 1676–1681.

Laan, L. A., v Haeringen, A., & Brouwer, O. F. (1999). Angelman syndrome: A review of clinical and genetic aspects. *Clinical Neurology and Neurosurgery*, *101*(3), 161–170.

Levenson, J. M., & Sweatt, J. D. (2005). Epigenetic mechanisms in memory formation. *Nature Review Neuroscience*, *6*(2), 108–118.

MacCabe, J. H., Lambe, M. P., Cnattingius, S. , Torrång, A. , Björk, C. , Sham, P. C. et. al. (2008). Scholastic achievement at age 16 and risk of schizophrenia and other psychoses: A national cohort study. *Psychology and Medicine, 38*, 1133–1140.

MacCabe, J. H., Leon, D.A., & Koupil, I. (2009) Lifetime reproductive output over two generations in patients with psychosis and their unaffected siblings: The Uppsala 1915–1929 Birth Cohort Multigenerational Study. *Psychology and Medicine,* doi:10.1017/S0033291709005431

Mefford, H. C., Sharp, A. J., Baker, C., Itsara, A., Jiang, Z., Buysse, K., et al. (2008). Recurrent rearrangements of chromosome 1q21.1 and variable pediatric phenotypes. *New England Journal of Medicine, 359,* 1685–1699.

Meyer-Lindenberg, A., Mervis, C. B., & Berman, K. F. (2006). Neural mechanisms in Williams syndrome: A unique window to genetic influences on cognition and behaviour. *Nature Review Neuroscience, 7*(5), 380–393.

Mill, J., & Petronis, A. (2007). Molecular studies of major depressive disorder: The epigenetic perspective. *Molecular Psychiatry, 12*(9), 799–814.

Mill, J., Tang, T., Kaminsky, Z., Khare, T., Yazdanpanah, S., Bouchard, L., et al. (2008). Epigenomic profiling reveals DNA-methylation changes associated with major psychosis. *American Journal of Human Genetics, 82*(3), 696–711.

Murphy, K. C., Jones, L. A., & Owen, M. J. (1999). High rates of schizophrenia in adults with velo-cardio-facial syndrome. *Archives of General Psychiatry, 56*(10), 940–945.

Penagarikano, O., Mulle, J. G., & Warren, S. T. (2007). The pathophysiology of fragile X syndrome. *Annual Review of Genomics and Human Genetics, 8,* 109–109.

Plomin, R., DeFries, J. C., McClearn, G. E., & McGuffin, P. (Eds.). (2008). *Behavioral genetics* (5th ed.). New York: Worth Publishers.

Redon, R., Ishikawa, S., Fitch, K. R., Feuk, L., Perry, G. H., Andrews, T. D., et al. (2006). Global variation in copy number in the human genome. *Nature, 444*(7118), 444–454.

Reichenberg, A., & Harvey, P. D. (2007). Neuropsychological impairments in schizophrenia: Integration of performance-based and brain imaging findings. *Psychological Bulletin, 133*(5), 833–858.

Reichenberg, A., Weiser, M., Rabinowitz, J., Caspi, A., Schmeidler, J., Mark, M., et al. (2002). A population-based cohort study of premorbid intellectual, language, and behavioral functioning in patients with schizophrenia, schizoaffective disorder, and nonpsychotic bipolar disorder. *American Journal of Psychiatry, 159*(12), 2027–2035.

Sebat, J., Lakshmi, B., Malhotra, D., Troge, J., Lese-Martin, C., Walsh, T., et al. (2007). Strong association of de novo copy number mutations with autism. *Science, 316*(5823), 445–449.

Spreen, O., Risser, A. H., & Edgell, D. (Eds.).(1995). *Developmental neuropsychology.* New York: Oxford University Press.

Stefansson, H., Rujescu, D., Cichon, S., Ingason, A., Steinberg, S., Fossdal, R., et al. (2008). Large recurrent microdeletions associated with schizophrenia. *Nature, 455*(7210), 232–236.

Stromme, P., Bjornstad, P. G., & Ramstad, K. (2002). Prevalence estimation of Williams syndrome. *Journal of Child Neurology, 17*(4), 269–271.

Tan, H. Y., Chen, Q., Sust, S., Buckholtz, J. W., Meyers, J. D., Egan, M. F., et al. (2007). Epistasis between catechol-O-methyltransferase and type II metabotropic glutamate receptor 3 genes on working memory brain function. *Proceedings of the National Academy of Sciences of the United States of America, 104*(30), 12536–12541.

Tsankova, N., Renthal, W., Kumar, A., & Nestler, E. J. (2007). Epigenetic regulation in psychiatric disorders. *Nature Review Neuroscience, 8*(5), 355–367.

Walsh, T., McClellan, J. M., McCarthy, S. E., Addington, A. M., Pierce, S. B., Cooper, G. M., et al. (2008). Rare structural variants disrupt multiple genes in neurodevelopmental pathways in schizophrenia. *Science, 320*(5875), 539–543.

Wang, P. P., Solot, C., Moss, E. M., Gerdes, M., McDonald-McGinn, D. M., Driscoll, D. A., et al. (1998). Developmental presentation of 22q11.2 deletion (DiGeorge/velocardiofacial syndrome). *Journal of Developmental & Behavioral Pediatrics, 19*(5), 342–345.

Weaving, L. S., Ellaway, C. J., Gecz, J., & Christodoulou, J. (2005). Rett syndrome: Clinical review and genetic update. *Journal of Medical Genetics, 42*(1), 1–7.

Weeber, E. J., Levenson, J. M., & Sweatt, J. D. (2002). Molecular genetics of human cognition. *Molecular Intervions, 2*(6), 376–391, 339.

Williams, H. J., Owen, M. J., & O'Donovan, M. C. (2007). Is COMT a susceptibility gene for schizophrenia? *Schizophrenia Bulletin, 33*(3), 635–641.

Williams, N. M., & Owen, M. J. (2004). Genetic abnormalities of chromosome 22 and the development of psychosis. *Current Psychiatry Report, 6*(3), 176–182.

Xu, B., Roos, J. L., Levy, S., van Rensburg, E. J., Gogos, J. A., & Karayiorgou, M. (2008). Strong association of de novo copy number mutations with sporadic schizophrenia. *Nature Genetics, 40*(7), 880–885.

COGNITIVE NEUROPSYCHIATRY
2009, 14 (4/5), 391–418

Psychology Press
Taylor & Francis Group

Genome-wide strategies for discovering genetic influences on cognition and cognitive disorders: Methodological considerations

Steven G. Potkin[1], Jessica A. Turner[1], Guia Guffanti[1,2], Anita Lakatos[1], Federica Torri[2], David B. Keator[1], and Fabio Macciardi[1,2]

[1]*Department of Psychiatry and Human Behavior, University California of Irvine, Irvine, CA, USA,* [2]*Department of Science & Biomedical Technologies – Fondazione Filarete and School of Medicine, Università degli Studi di Milano, Milan, Italy*

Introduction. Genes play a well-documented role in determining normal cognitive function. This paper focuses on reviewing strategies for the identification of common genetic variation in genes that modulate normal and abnormal cognition with a genome-wide association scan (GWAS). GWASs make it possible to survey the entire genome to discover important but unanticipated genetic influences.
Methods. The use of a quantitative phenotype in combination with a GWAS provides many advantages over a case-control design, both in power and in physiological understanding of the underlying cognitive processes. We review the major features of this approach, and show how, using a General Linear Model method, the contribution of each Single Nucleotide Polymorphism (SNP) to the phenotype is determined, and adjustments then made for multiple tests. An example of the strategy is presented, in which fMRI measures of cortical inefficiency while performing a working memory task are used as the quantitative phenotype. We estimate power under different effect sizes (10–30%) and variations in allelic frequency for a Quantitative Trait (QT) (10–20%), and compare them to a case-control design with an Odds Ratio (OR) of 1.5, showing how a QT approach is superior to a traditional case-control. In the presented example, this method identifies putative susceptibility genes for schizophrenia which affect prefrontal efficiency and have functions related to cell migration, forebrain development and stress response.
Conclusion. The use of QT as phenotypes provide increased statistical power over categorical association approaches and when combined with a GWAS creates a

Correspondence should be addressed to Steven G. Potkin, Department of Psychiatry and Human Behavior, 5251 California Ave, Suite 240, Irvine, CA 92617, USA. E-mail: sgpotkin@uci.edu

http://www.psypress.com/cogneuropsychiatry DOI: 10.1080/13546800903059829

strategy for identification of unanticipated genes that modulate cognitive processes and cognitive disorders.

Keywords: Cognition; Imaging phenotype; Endophenotype; Quantitative trait; GWAS; Power; Permutation; Replication.

INTRODUCTION

Genes play a well-documented role in determining normal cognitive abilities as well as the development of cognitive disorders. This paper focuses on strategies for the identification of normal genetic variation in genes that modulate normal and abnormal cognition by combining brain images acquired during cognitive performance with genetic data. This approach is part of an emerging field referred to as Imaging Genetics. There are several excellent comprehensive reviews on Genome-Wide Association Study (GWAS) design methodology (Cardon & Bell, 2001; Hirschhorn & Daly, 2005; McCarthy et al., 2008; Thomas, Haile, & Duggan, 2005; Zondervan & Cardon, 2004) and corresponding statistical issues (Balding, 2006; Ziegler, Konig, & Thompson, 2008), as well as stratification (Cardon & Palmer, 2003; Freedman et al., 2004; Price et al., 2006; Serre et al., 2008), replication (Chanock et al., 2007) and epistasis (Keedwell et al., 2008; Phillips, 2008) that this paper refers to and builds upon. The reader is referred to these publications for greater depth on these subjects.

Imaging genetics leverages the strengths of both neuroimaging and genetic studies, visualizing the brain activation patterns in the context of genetic background. Imaging as an intermediate phenotype can clarify the functional link between genes and cognition. While imaging studies reveal many functional (and structural) aspects of function and dysfunction in cognition, their explanatory power may be limited by not considering the genetic bases of cognition, as many aspects of cognition are clearly heritable (Kennedy, Farrer, Andreasen, Mayeux, & St George-Hyslop, 2003). Cognitive performance may be more distant from the biological mechanisms underlying cognition than are brain images of functional circuitry (Basile, Masellis, Potkin, & Kennedy, 2002; Potkin, Anand, Alphs, & Fleming, 2003; Potkin, Kennedy, & Basile, 2002a, 2002b), as cognitive performance is a complex expression of function of several brain regions working as coordinated brain circuits. Given the known importance of both genetics and environment in brain function, and the role of neuroimaging in revealing brain dysfunction, the synergy of integrating genetics with brain imaging carries clear advantages:

- the function of the genes of interest can be revealed in neuroimaging data, an intermediate phenotype which has relevance to normal

cognition, as well as to many diseases in which cognition is abnormal (e.g., Meyer-Lindenberg & Weinberger, 2006);
- the unwanted variation in imaging studies can be reduced through controlling for genetic background;
- neuroimaging may identify dimensions of cognitive functions that are more closely related to susceptibility genes than are current clinical cognitive standardized assessments (e.g., Gottesman & Gould, 2003);
- gene discovery – identification of new "candidate" genes related to normal and abnormal cognition that would not be discovered by traditional candidate gene approaches (e.g., Papassotiropoulos et al., 2006; Potkin et al., 2009b,c).

The availability of high-throughput genotyping technologies and genomic resources such as HapMap (www.hapmap.org) has made it possible to survey the entire genome and increase the probability of discovering important unanticipated genetic influences. In this paper, we argue that the use of Quantitative Traits (QT) as phenotypes provide increased statistical power over categorical association approaches, and when combined with a GWAS creates a strategy for identification of genes that modulate cognitive processes.

The outline of this paper is as follows: first we review the motivation for performing a GWAS, followed by a short review of what constitutes a GWAS and the major statistical issues involved in its analysis. Then we discuss the use of cognitive measures as phenotypes in GWAS, the increase in power involved with such use, and present our General Linear Model (GLM) approach, with a fuller discussion of the statistical considerations and the resulting interpretation that follows from this approach, using an imaging genetics example from previously published work.

MOTIVATION FOR PERFORMING A GWAS

Much has been learned from candidate gene approaches, in which *a priori* hypotheses are tested. These hypotheses may be based on linkage studies that have identified a key region of the genome as linked to a cognitive function, or based on current physiological or pathophysiological knowledge. This approach is very attractive as it takes current knowledge, and extends it. However, a limitation of this approach is precisely that it starts with our current understanding of physiology or pathophysiology, which we know is incomplete.

For example, both acetylcholine and dopamine are involved in cognitive function (Gray & Roth, 2007; Robbins & Roberts, 2007; Tamminga, 2006), and investigation of the synthesis/release/re-uptake and metabolism of these neurotransmitters are leads to a number of obvious candidate genes targets. What is less clear is whether and how these monoamines are modulated or regulated by various transcription factors, or other regulatory elements or growth factors, such as the Brain Derived Neurotrophic Factor (BDNF). BDNF could then be considered as a candidate gene. Such regulatory influences can be determined to some extent by investigating *in silico* pathway and epigenetic databases. Still, there are many important interactions and functions not yet identified. Additionally, this process typically leads to one study per candidate gene, and, with thousands of genes expressed in the brain to consider, is inefficient. While such interactions are continuing to be discovered, we do not yet know how to place the dopaminergic system in context within the whole brain or genetic milieu.

It is also important to keep in mind that most of the current cognitive candidate genes have only become candidates through *post hoc* genetic annotation and selection (Abdolmaleky, Faraone, Glatt, & Tsuang, 2004; Egan et al., 2001; Lohmueller, Pearce, Pike, Lander, & Hirschhorn, 2003; Shifman et al., 2002). Plausible hypotheses build upon current knowledge will extend that knowledge (Meyer-Lindenberg et al., 2006; Roffman, Weiss, Goff, Rauch, & Weinberger, 2006), but are unlikely to result in a fundamental shift in the conceptualization of cognition.

An alternative approach is to screen the entire genome (GWAS) for genes that influence these cognitive processes or disorders, without being limited to potentially less important chromosomal regions by our current state of incomplete knowledge. This strategy applied to Imaging Genetics can identify new, unanticipated candidate genes for cognition. These insights have the potential to fundamentally alter our current understanding of cognitive processes, since this application of Imaging Genetics can visualize the brain activation patterns in the context of genetic background, thereby synergizing the strengths of each individual approach (Potkin et al., 2009b,c) and ultimately representing a strategy for gene discovery.

WHAT CONSTITUTES A GENOME WIDE ASSOCIATION STUDY

A GWAS surveys most of the genome for potentially causally-related Single Nucleotide Polymorphisms (SNPs). The selection criterion of the markers to include in a GWAS is based on the hypothesis and methods of Linkage Disequilibrium (LD) (Lewontin, 1958) between genetic marker alleles and the true causative alleles (Durrant et al., 2004). The two most common statistical measures used to summarize LD between two markers are D' and r^2. Both measures are built on the difference between the probabilities of observing two markers (e.g., SNPs) alleles on the same haplotype compared to observing them independently in the population. The D' statistics is equal to 1 unless recombination across the two SNPs has occurred, therefore the higher the D' the greater the LD; however, the allele frequency of the SNPs is not considered. The r^2 measure directly depends on allele frequency; a r^2 value of 1 can be achieved only when the marker loci have identical allele frequencies and every occurrence of each allele marker perfectly predicts the allele at the other locus (i.e., an identical proxy). An r^2 of 1 thus indicates two SNPs are in perfect LD so that only one SNP needs to be genotyped to know the genotype of the other. This "tag" SNP approach can provide a high level of genomic coverage as a large number of untyped SNPs will be identifiable with this reference set of tagging-SNPs genotyped. To generate a comprehensive LD map of the human genome, the HapMap Project has already genotyped the most common DNA variants and is now completing the identification of less frequent variants (Kruglyak, 2008). The general consensus is that an r^2 of .8 or greater is sufficient for a tag SNP to obtain a good coverage of LD-dependent untyped SNPs. Thus tagging SNPs allow genotyping of far fewer marker SNPs with relatively small losses in power (Anand et al., 2005). The tagging SNP method is effective even with alleles frequency $< 5\%$ and therefore even rare variants can be detectable (Fallin et al., 2001).

The completion of the human genome sequence and advances in SNP genotyping have made this kind of coverage both feasible and affordable (Botstein & Risch, 2003; Hirschhorn & Daly, 2005; Stranger et al., 2005). Current microarray development, like for example the Illumina Human1M-Duo BeadChip (Di Piero, Ferracuti, Sabatini, & Pantano, 1994) currently covers 96% (in CEU) of the genome with 1 million SNPs at average LD-related distance corresponding to an r^2 of .8. The costs of GWAS have also decreased to the point that they are being almost routinely conducted for many disorders even in medium size laboratories.

STATISTICAL CHALLENGES IN GWAS

GWAS techniques produce anywhere from 100,000 to more than 1,000,000 genotypes, e.g., data points, per subject. Classical statistical analytical techniques are not designed for situations where the number of variables so grossly outnumbers the number of subjects, and such analyses have rarely been attempted in biological studies. The classical approach of considering each association between a SNP and the phenotype as a "repeated" test, and thus correcting for the number of tests performed (under the assumption that they are multiple tests), is out of its depth when faced with a million or more tests. The potential for detecting genetic influences on phenotypes is lost if the sample size required for study by conventional measures literally verges on a significant proportion of the world's population. The technical developments in genotyping have apparently outstripped the ability to analyze the resulting information, if we keep using statistical methods and strategies that have been developed for much simpler tasks. In fact, the traditional methods to correct for multiple tests, for example the well-known Bonferroni's method is not ideal in this application, since it assumes that all the factors, the SNPs in our case, are independent from each other. However, this is not the reality of the genome, where SNPs are somewhat correlated as function of their relative distance independently from their status as "taggers" or not.

While new methods have been developed (Benjamini & Hochberg, 1995) to address this problem, a more compelling theoretical question is whether we are dealing with a problem of multiple testing or rather with an issue of multiple hypotheses. In fact, if we consider each single SNP a possible "cause" of the disease that we are investigating, then testing one million SNPs reverts to testing one million different hypotheses, each one of them with a high degree of uncertainty due to the overall complexity of the genome in addition to the number of hypotheses tested. As one attempts to deal with the "one million hypotheses" question, the proposed solutions span a wide range of alternatives: at one extreme, theoretical epidemiologists (Rothman, 1990) state that there is no need to correct and the investigator should simply report any significant finding together with the number of hypotheses tested, leaving the burden to a subsequent focused experiment to prove or disprove the previous reported positive findings.

With a more conservative view, others focus on the risk of too many potential false positives and attempt to balance and trade-off between the complementary risks of false positives and false negatives. They propose to minimize the risk of false positives with techniques such as permutations (Westfall, 1993), FDR (Benjamini & Hochberg, 1995) and others. This latter approach is more or less conceptually based on *a priori* establishing a threshold for a "genome-wide" significant finding, almost in the way R.A. Fisher originally set up the 5% significant threshold for classical statistical tests (Fisher, 1935). It is important to note that false negatives can be potentially more harmful than false positives, since may be discarded forever, while a false positive can always be disproved or "falsifiable" albeit at some cost. Any established threshold for genome-wide significance is a subjective decision that cannot be universally proven or disproven, regardless of the arguments in support of a given threshold.

In addition to the problem of multiple testing/multiple hypotheses, there are several other important issues that are the current focus of interest in statistical genetics. Other topics that are beyond the scope of the present paper, include research related to how to statistically identify "genes" (or chromosomal regions) rather than SNPs, given that our primary interest is in mapping putative functional elements of the genome rather than simple DNA point variations. Addressing gene–gene interactions (Brzustowicz, 2008; Moore, 2008), hence epistasis (Chapman & Clayton, 2007; Evans, Marchini, Morris, & Cardon, 2006; Jiang, Tang, Wu, & Fu, 2009) for broad, or gene–environment (Clayton & McKeigue, 2001; Glazier, Nadeau, & Aitman, 2002; Hoffmann, Lange, Vansteelandt, & Laird, 2009; Lander & Kruglyak, 1995) interactions or even how to validate a causative or regulatory network (Barabasi, 2007; Hidalgo, Blumm, Barabasi, & Christakis, 2009) are areas of active statistical research.

It is in fact becoming increasingly evident that multivariate analysis methods can be applied to extract meaningful results from a GWAS, including the complex interdependence that exists across the SNPs within the genome. These multivariate approaches not only address a gap in our current analytical strategies, but can reach new and sometimes unexpected information in the way our genome is organized and how it can be deranged in complex disease (Slawski, Daumer, & Boulesteix, 2008). The application of these methods such as Random Forest, Support Vector Machines, etc. is finding increased popularity as classification techniques in gene expression datasets for identifying risk genes, or to case-control GWS analyses (for an introduction to their applications in gene expression analyses see, Slawski, Daumer, & Boulesteix, 2008). While application of some of these methods to quantitative regressions are possible, they have been applied to QT with less frequency (though see Calhoun, Liu, & Adali, 2009) for ICA analyses of imaging and genetics data, and Lee, Sun, Kustra, & Bull, 2008, for an example of applying Random Forest techniques with the QT of blood pressure in a linkage study). Their application to QT such as cognitive phenotypes in GWS is still very much in development. Some of the issues that are challenges for the univariate analyses do not apply in the multivariate analyses, e.g., repeated testing of hypotheses, but the larger issues of interpretation and confirmation of the results apply in either analytical approach. We focus in this paper on the application of massively univariate statistical techniques in imaging genetics analyses.

COGNITION AS A PHENOTYPE

GWAS have also found genetic influences on various cognitive measures. One of the most studied traits is human memory, with a heritability estimate of approximately 50% for various measures (McClearn et al., 1997). Performance on a memory task has been used successfully used as a QT in a GWAS study of young healthy adults, leading to the identification of KIBRA as a memory-related gene (Papassotiropoulos et al., 2006). A common variant of the KIBRA gene, identified by the SNP rs17070145, was found to be significantly associated with memory performances, based on quartile ranking in verbal episodic memory, in three independent samples. The genome significance threshold for the first GWAS was established based on a strategy that combined two different statistical approaches: a single-point method using Bonferroni's correction for independent comparisons and a sliding-window approach to identify chromosomal clusters harboring the most significant SNPs at various window sizes. The positive findings selected for high "statistical confidence" were further genotyped in independent samples (Papassotiropoulos et al., 2006). Subsequently, non-carriers of the T allele (rs17070145) in KIBRA were found to be associated with an increased risk of developing late-onset Alzheimer's disease (AD) (Corneveaux et al., in press). KIBRA was over-expressed and three of its four known binding partners under-expressed in hippocampal, posterior cingulate, and temporal cortex regions of AD subjects studied post mortem. Interestingly, cognitively normal, late-middle-aged persons who were non-carriers of the T allele exhibited lower glucose metabolism on PET scanning in posterior cingulate and precuneus, brain regions sensitive to detecting AD, than did carriers. These findings suggest that KIBRA is associated with individual variation in episodic memory in healthy normals as well as in the predisposition to AD. It is noteworthy that the case-control study required far more subjects to achieve the same statistical power as that found in the much smaller study using cognition as a QT.

Almasy et al. (2008) used a QT design to examine multiple cognitive domains in families with schizophrenia, using 386 autosomal microsatellite markers (selected from the Genethon human linkage map) in a genome-wide linkage study. In this study they identified evidence for genetic loci on chromosome 19 and chromosome 5q.31–34 associated with cognitive performance, as well as with risk for schizophrenia. The Papassotiropoulos and Almasy studies indicate the value of employing QT where unanticipated genes are identified and become candidates for further molecular and functional exploration. Their work expands the potential mechanisms and circuitry underlying cognitive function and provides new scientific models of brain function, as well as new targets for therapeutic interventions.

POWER OF A QUANTITATIVE TRAIT ANALYSIS IN A GWAS

The power of an association study depends on the specific design of the study, such as whether replication or joint analyses are to be accommodated. Several methods have been developed to minimize the financial costs of genotyping, such as a two-stage design (Ennis et al., 2008; Li, 2007; Skol, Scott, Abecasis, & Boehnke, 2007). These designs recognize the need to balance the risk of false positives with the greater need to minimize false negative results. False negative results may be more of a problem, as false positive results will be identified in joint analysis studies or properly powered replication studies. In any event, an appropriate correction for multiple comparisons/multiple hypotheses must be used, but determining the appropriate correction is challenging for GWAS studies as we and others have discussed (Balding, 2006; Thomas, 2006; Ziegler et al., 2008).

Our approach to power analysis is different from that assumed in standard association studies since we will focus on quantitative phenotypes as contrasted with case controls. We estimate power under different assumptions regarding the effect size and the variation in allelic frequency, using the approach proposed by Purcell, Cherny, and Sham (2003) for a quantitative complex trait. The results are presented in Figures 1 and 2. For comparison, power curves using a categorical, case-control design are also shown with an odds ratio (OR) of 1.5. The usual cautions apply here: for example, the putative allelic frequency of the "cognitive or disease" allele should not be too different from the "marker" allele frequency, with the risk of missing the association (Durrant et al., 2004). The following conservative assumptions, an effect size of 10%, a low minor allele frequency of 20% in Figure 1 and 10% in Figure 2, a sample size of 800 and 500, respectively, will provide 80% power for a QT phenotype to be detected even when the estimated p-value is very conservative ($p \leq 10^{-7}$), for both a multiplicative

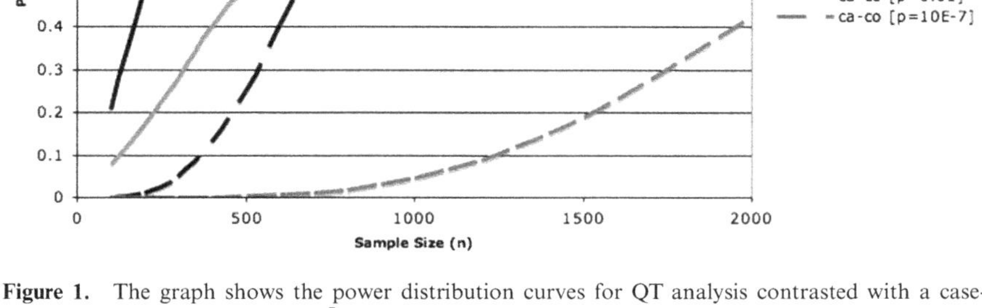

Figure 1. The graph shows the power distribution curves for QT analysis contrasted with a case-control design at $p < .01$ and 10^{-7} (OR = 1.5). The x axis portrays the sample sizes and the y axis the power at each value of the sample size for a 10% percentage variance explained for the QT, a QT MAF of 10% and a marker SNP MAF at 20%. The results for QT and case-control are displayed in black and grey respectively. Compared to QT, the case-control curves are shifted to the right indicating that much larger sample sizes are required to reach the same power.

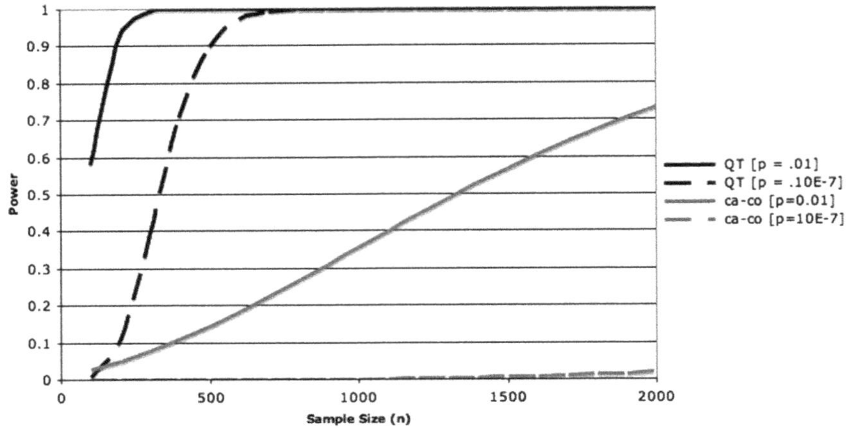

Figure 2. The graph shows the power distribution curves for QT analysis, contrasted with a case-control design using the same parameters as in Figure 1 but with the marker SNP MAF at 10%. Comparison between Figures 1 and 2 highlights the effects of a satisfactory match between tagging and causal SNP MAFs. See Figure 3 for more details.

and an additive model. In contrast, a much larger sample of many thousands is required to obtain 80% power in a case-control design, given an OR of 1.5 or slightly greater.

Table 1 shows, for a constant 80% power, how the needed sample size to detect a related locus with a QT phenotype depends on the amount of variance explained in the QT and in the SNP allele frequencies, over a 10–30% range. A range of minor allele frequencies (MAFs) for the tagging SNPs is also indicated, assuming a stable and constant LD (measured as D') between the tagging SNP and the causative SNP of .95. As the variance of the QT, explained by a given SNP, increases, the sample size dramatically decreases for all frequencies of the tagging SNPs. The curve distributions in Figures 1–3 again indicate that the closer the match between the allelic frequency of the QT locus and the tagging SNP, the greater the power and the smaller the needed sample size, while the relationship is very different in the case-control approaches. It is well-known that our ability to detect a "causative" SNP in a case-control design is also a function of how close the allelic frequencies are for the unobserved SNP and the marker or tagging-SNP even when the allelic frequency for the causative SNP is as high as 30%. If we cannot match the causative SNP frequency with the "tagging" SNPs that we are experimentally testing, we risk missing an important association unless we use an unrealistically large sample size. Figure 3 shows how the power can vary as a function of the mismatch between the two allelic frequencies of the causative and tagging SNPs, even under the assumption of high LD between them.

TABLE 1
Sample sizes needed to reach 80% power to detect a locus responsible for the QT

	QT MAF=.1			*QT MAF=.2*			*QT MAF=.3*		
SNP MAF	*QTL .10*	*QTL .20*	*QTL .30*	*QTL .10*	*QTL .20*	*QTL .30*	*QTL .10*	*QTL .20*	*QTL .30*
0.05	**961**	**470**	**146**	1064	528	349	3957	1968	1305
0.1	**439**	**209**	**132**	1041	510	333	1847	913	602
0.2	1026	502	328	**440**	**210**	**132**	793	386	250
0.3	1781	880	580	780	380	247	**441**	**210**	**132**
0.4	2787	1383	915	1241	610	400	**714**	**346**	**224**
0.5	4195	2087	1385	1882	931	613	**1097**	**538**	**351**

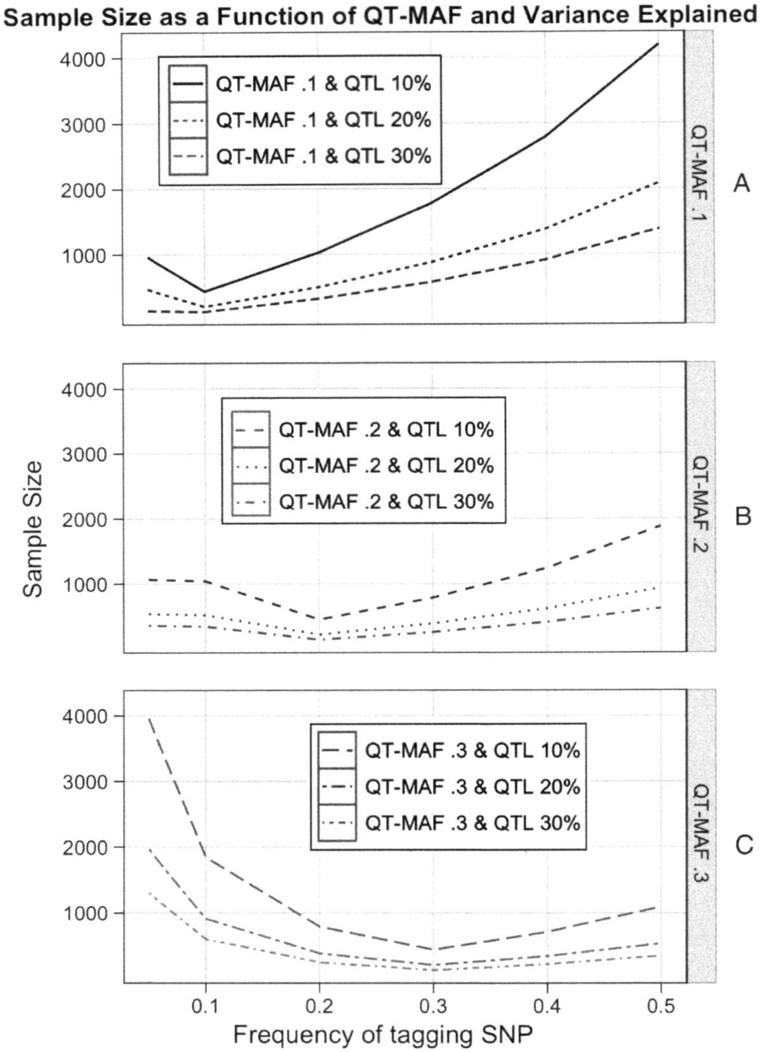

Figure 3. The graph depicts the sample size required for a power = 80% for a QT phenotype with tagging SNP frequencies from .05 to 0.5 given a 0.10 (panel A), 0.20 (panel B), and 0.30 (panel C) MAF for the QT and when the total variance explained by the QT is 10, 20 and 30%. The curves show the sample size required for combinations of the QT values: the lowest sample size is reached when there is agreement between the allelic frequencies of the tagging SNPs with that of the actual QT allele, which is unknown. This is accurate, independent of the total amount of variance explained by the QT.

STATISTICAL MODEL FOR A QT PHENOTYPE

All SNPs that pass quality control checks (Teo, 2008) are included in the GWAS analysis. The simplified statistical model compares the differential effects of SNP alleles or genotypes on the QT.

The simplest GLM is:

$$Y = \mu + \beta_1 {*}SNP + \beta_i {*}V_i + \varepsilon \qquad (1)$$

in which Y is the QT, μ is the population mean, β_i are the coefficients being estimated, *SNP* represents the marker being tested, V_i are other variables in the model (e.g., the age of the subject, diagnostic group, gender, etc.), and ε is the error term. The SNP term represents the main effect of the "gene" on the QT, independent of any class grouping or other variable.

The beauty of the GLM approach lies in its extensibility. In the study of a cognitive QT in the presence of a cognitive disorder (e.g., looking for genetic effects modulating both the development of autism and specific cognitive dysfunction within the disease), the GLM can be extended to:

$$Y = \mu + \beta_1 {*}SNP + \beta_2 {*}diagnosis + \beta_3 {*}SNP{*}diagnosis + \beta_i {*}V_i + \varepsilon \qquad (2)$$

In which the $\beta_2{*}diagnosis$ term indicates how the cognitive phenotype Y differs across the diagnostic variable. The SNP term represents the overall effect of genotype on the QT; the SNP*diagnosis interaction term in this model is an integrative term reflecting how the genotype differentially affects cognitive phenotype in the disorder. This approach can determine the effects of genes on cognition as measured by brain activation or behaviorally measured performance, and further, it can determine if these genetic effects differ by diagnosis.

Depending on the study design, of course, the GLM can also be extended to explicitly model effects of known, candidate genes, and/or gene–gene interactions. In the example below, Y could be a quantitative measure of attention in a study of AD and healthy subjects with different APOE alleles, and the research question is looking for genetic effects on cognitive function that interact with the APOE genotype:

$$Y = \mu + \beta_1 {*}SNP + \beta_2 {*}diagnosis + \beta_3 {*}SNP{*}diagnosis + \beta_4 {*}APOE + \beta_5 {*}APOE{*}SNP + \beta_i {*}V_i + \varepsilon \qquad (3)$$

How complicated the analysis model can be will depend, of course, on other design considerations such as expected effect sizes and sample sizes.

Covariates in the model

When there is an established strong genetic influence on the phenotype, it is important to eliminate or control for this known genetic effect to enhance the strategy to identify unanticipated genes. For example, APOE4 is a well documented risk factor for AD, and thus should be controlled or covaried for, as the effects of APOE4 can otherwise obscure the identification of other risk genes. Other relevant variables can also be included in the model. In the case of AD, we can include age and gender as covariates in the regression model as they are associated with the risk of developing AD.

GENETIC MODELS

The genetics underlying complex traits is characterized by a sophisticated interplay between multiple genes. When applied to common diseases, genome wide association studies must take into account the degree of etiopathological complexity using the most appropriate model. Classical statistical genetics offers four possible models to code the SNP term in the GLM above: *additive, codominant, dominant* and *recessive*. The *additive* model is generally preferred as it reflects the additive contribution to risks for complex diseases. Additive models can also detect strong non-additive effects (Gianola & de los Campos, 2008; Hill, Goddard, & Visscher, 2008; Lee, van der Werf, Hayes, Goddard, & Visscher, 2008; Yamada & Okada, 2009). When the hypothesized risk allele (B) is rare, the dominant model can be used as it pools the homozygotes (BB) and heterozygotes (AB) genotypes together in the analysis. Such a model tests the hypothesis that carrying even one copy of that particular allele increases the risk of disease.

This approach may not be exhaustive when applied to complex traits, as these models were developed to study single major loci. For complex traits, functional pathways should be considered because multiple genes are involved and interact. Since additive, dominant, codominant and recessive models refer to the single SNP as the unit of the analysis, new statistical approaches that model interaction effects, either gene–environment (e.g., gene × diagnosis) and gene–gene (i.e., epistasis) are needed. There is relatively little work on how to model these interaction relationships, and it is an area of active research (Moore, 2003; Phillips, 2008; Wang, Comaniciu, & Fasulo, 2006).

HOW TO CALCULATE A GWAS THRESHOLD

A major challenge to the interpretation of results from a GWAS is setting the appropriate statistical threshold. In classical statistics, as the number of tests against a single null hypothesis increases, the statistical threshold (*p*-value) has to account for the probability of a false positive occurring by chance. This is typically done using a Family-Wise Error Rate (FWER) approach such as a Bonferroni correction; if a GWAS is going to test for

1 million SNPs (and not considering here the complexities potentially related to testing 1 million hypotheses that we have previously discussed), the set significance level is 10^{-8} for a global significance level of 5%. However, the Bonferroni's correction is not appropriate for observational studies such as GWAS (Perneger, 1998) because it does not account for the dependencies of SNPs that are close to each other across chromosomes, thus leading to an overcorrection. Arking et al. (2006) proposed less stringent significance levels that accounted for LD across SNPs. Zondervan and Cardon (2007) provide a method to adjust for the actual number of independent tests (Zondervan & Cardon, 2007). The HapMap consortium proposed a local significance threshold of 5.5×10^{-8} based on re-sampling from empirical data under the null hypothesis (Altshuler et al., 2005), reaching a similar result to the WTCCC consortium (5×10^{-7}) (Dudbridge & Gusnanto, 2008; WTCCC, 2007).

Various combinations of a classical method with other methods have been suggested, including multi-stage experimental designs, ranking the results and identifying a top subset of likely genes, and permutation testing to determine an empirical *p*-value. Also, expanding the work by Benjamini and Hochberg (1995), Efron and Tibshirani (2002) proposed controlling for the risk of false positives over all the positive results rather than over all the possible tests (the False Discovery Rate approach – FDR), thus partially compensating for the overcorrection of the more traditional FWER methods.

It is important to re-emphasize that all these proposed methods address the evaluation of a single SNP at a time. If two or more SNPs are in (strong) LD to each other and show a similar pattern of significant association with the trait of interest, it is appropriate to use a less stringent threshold (Ziegler et al., 2008). Alternatively one can avoid putting forth a prespecified formal significance threshold, and present all results ordered from lowest to highest *p*-values (Almasy et al., 2008; Helgadottir et al., 2007).

In GWAS, a false positive results in an additional cost of following up the initial finding, always a necessity considering the exploratory nature of GWAS. The consequences of false positives in a GWAS are less severe than finding a false positive in a pivotal clinical trial that could lead to potentially dangerous treatment (Rothman, 1998). Importantly, any correction for multiple testing to avoid increasing false positives also affects the false negative rate, reducing power to detect a true significant finding (Samani et al., 2007). The latter may be a more serious error, in that an important finding is prematurely dismissed.

Several other approaches are being developed to determine which GWAS significance threshold is most appropriate to a specific research question. It is important to keep in mind that in a GWAS each SNP is its own hypothesis; a GWAS involves testing of hundreds of thousands of hypotheses. This is a fundamentally different question than testing the same hypothesis a million times. Bonferroni corrections are more suited to the latter case but are not well suited to the testing of many different hypotheses (WTCCC, 2007). Perneger (1998) made the case that Bonferroni adjustments are concerned with the wrong hypothesis (i.e., that all null hypotheses are true simultaneously, which is not of interest) and increases the likelihood of type II errors concluding that the Bonferroni method may create more problems than it solves. Appropriate statistical correction for multiple statistical tests is an area of intense statistical research (Dudoit & van der Laan, 2008). Despite a lack of consensus various practical approaches are in use to address the problem of multiple testing and determining the appropriate GWAS threshold, several of which are briefly summarized.

Permutation tests

Permutation methods (e.g., Manly, 1997) offer the possibility of using the dataset collected to empirically determine the statistical threshold, both for case-control and QT phenotypes. Although computationally expensive, current technology is robust enough to perform permutation testing.

While permutation methods are a broad class of techniques, in GWAS applications they are used to determine the proportion of cases in which the *F* or chi-square statistics would arise under the null hypothesis of no genetic effect. With a permutation approach, the "labels" identifying cases and controls (or a given measure for a quantitative phenotype) are randomly reassigned, the original analysis re-done and the subsequent statistics is noted; when this is done many thousands or hundreds of thousands of times, the distribution of the test statistic under the null hypothesis for that specific dataset is known. This data derived distribution can deviate significantly from the *a priori* distribution. The probability of the original statistics in the current sample arising by chance, then, can be empirically estimated.

In a dichotomous case-control study, when testing the main genetic effect on case-control status, usually the case-control designation is permuted across subjects; these methods are now standard with GWAS software (e.g., PLINK release v1·03, Purcell et al., 2007). However, for QT interactions it is not clear if the QT measure (*Y*) should be permuted, and what level of randomization (full, reduced, or constrained) is required. If the significance of an interaction term between SNP and diagnosis is being assessed, for example, the method may permute residuals for that term rather than the original raw data. Depending on the precise factorial design, full permutation may not be used, but rather permutation within exchangeable units in the design (for more detail on these issues, see for example, Anderson & Ter Braak, 2002; Jung, Jhun, & Song, 2006; Manly, 1997).

With increasing interest in identifying improved multiple testing procedures for GWAS, we suggest a possible strategy to modify the existing non-parametric inferential procedures in order to fit the interaction statistical assumptions. For illustrative purposes we propose to randomly select a sample of SNPs (e.g., say 1% or 5,000 of the 500,000 SNPs in the GWAS) and perform the permutation tests 1,000 times for each of the 5,000 SNPs creating a set of 5 million interaction p values, representing a null distribution for a random sample of SNPs in the GWAS. From this null distribution, the top 5 or 1% empirical threshold p-value can be set as an empirically determined, genome-wide threshold. The number of SNPs and permutations required to achieve the desired power will depend on sample size. This combines the strength of the permutation testing method with the GWAS data to determine an underlying null distribution for the full GWAS (see Weinberger presentation to Psych Gen NYC, 2008).

Additional SNPs in the same region

Combinations of SNPs can also be considered, through various "haplotype-based" approaches (Seasholtz et al., 2006), though that has not been applied to a GWAS with a QT. The greater the number of significant tagging SNPs in proximity to one another, the smaller is the likelihood of a finding being a false positive. As chromosomal regions rather than individual SNPs represent the inherited units into which an inherited genome can be partitioned, finding several significant tagging SNPs in physical proximity provide additional support for the locus being a true risk factor. The focus on tagging SNPs is critical, since we already know that adjacent SNPs are more likely to be in greater LD, and hence finding many close SNPs significant may not be particularly relevant since they share the same meaning. Haplotype approaches may also be able to capture effects in SNPs which are rare, in the 1–5% range (de Bakker et al., 2005; Kamatani et al., 2004; Lin, Chakravarti, & Cutler, 2004). The haplotype approach known as the "sliding window" (Fallin & Schork, 2000) can be used to detect the effects of rare alleles, using the extended block method of Purcell et al. (2007). A complete catalogue of all the low frequency SNPs and CNVs discovered will improve pinpointing genes, structural variants in chromosomes and other individual genomic variations that are associated with disease.

RESULTS OF AN IMAGING GENETICS ANALYSIS OF COGNITIVE FUNCTION

In our own studies, we have used functional neuroimaging as a quantitative phenotype in the context of GWAS to identify genes related to cognition in schizophrenia, and genes responsible for cognitive differences between subjects with schizophrenia and controls. We measured Blood Oxygenation Level Dependent (BOLD) signal in the dorsal lateral prefrontal cortex (DLPFC) during the Sternberg Item Recognition Paradigm (SIRP), a working memory task, in a group of schizophrenia and healthy control subjects (see Potkin et al., 2009b,c). In a similar fMRI study of schizophrenia and healthy control twins, DLPFC activation during the SIRP as well as the behavioral measures was found to be heritable (Karlsgodt et al., 2007).

DLPFC has a well documented role in memory performance in normal cognition as well as being a locus of dysfunction in schizophrenia; differences in the DLPFC between schizophrenic and control subjects have been described in local gene expression (Dean, Keriakous, Scarr, & Thomas, 2007; Vawter et al., 2004), cell morphometry, structural circuitry (Lewis & Gonzalez-Burgos, 2008; Shenton, Dickey, Frumin, & McCarley, 2001), and both local and distributed functional activation (Manoach et al., 1999; Menon, Anagnoson, Mathalon, Glover, & Pfefferbaum, 2001; Weinberger, Berman, & Zec, 1986).

In these data, schizophrenics show more BOLD activation in the DLPFC than do healthy controls when matched for accuracy performance on the Sternberg Item Recognition Paradigm (SIRP), a working memory task, consistent with cortical inefficiency (Potkin et al., 2009a). Subjects performed the Sternberg Item Recognition Paradigm (SIRP) with memory loads of 1, 3, or 5 items. During the "encode" condition, subjects memorized a set of target digits (memory load 1, 3 and 5 digits). They were then presented with probes (single digits) and responded by indicating whether the probe was a target (a member of the memorized set). The associated fMRI BOLD measure from the DLPFC while performing the probe condition and holding three items in memory was used as a quantitative phenotype in a GWAS.

Subjects were genotyped using the Illumina HumanHap370-Duo, providing 370404 SNPs suitable for analysis with the fMRI QT. Using the GLM analyses presented and statistical thresholds described above, two genes with functions related to cell migration to forebrain structures were identified, that had not been previously associated with cognition or schizophrenia (Potkin et al., 2009b). Six additional genes (or chromosomal regions) related to forebrain development and stress response, affecting prefrontal efficiency were also identified (Potkin et al., 2009c). These studies highlight the ability of brain imaging as a QT to identify unanticipated genes related to cognition.

INTERPRETATION OF THE RESULTS OF GWAS

The complexity of human psychiatric diseases and the questionable accuracy of the psychiatric diagnoses require a partitioning process of the potential underlying pathophysiological mechanisms (Meyer-Lindenberg & Weinberger, 2006). Imaging offers a tool that reflects underlying QT that may be more directly linked to the neurobiology of normal and abnormal function, serving as a proxy for the underlying causal mechanism. Therefore imaging phenotypes have great potential for gene discovery by increasing the power of a study and decreasing phenotypic heterogeneity (Walters & Owen, 2007). However, in choosing the imaging phenotype, choice of the task is an important consideration. A complex cognitive task, which may reflect important cognitive differences between groups, may be so complex that it is difficult to deconstruct into the many components and processes involved. On the other hand, a simple cognitive task, which is easy to interpret, may be so simple or narrow as to no longer reflect the essence or complexity of the disorder.

In addition, the combination of the QT with a whole genome wide scan can discover unanticipated genes previously not associated with the trait or the disorder. The discovery of novel associations between genes and risk for cognitive or neuropsychiatric disorder offers a powerful impetus to postulate new biological mechanisms. Although this approach is focused on discovery of novel genetic associations with cognitive function, the biological meaning of the associations emerging from the analyses must be considered.

The strategy leverages the collective biological insights of ancillary studies, published results from other research programs, databases, and genetic annotation. The consistency of findings across the different studies, and the correspondence to any *in vitro* functional experimental results based on the findings will determine how compelling the results are.

The annotation and comparative analysis of genes of interest can be a challenging and time consuming task. Fortunately, numerous databases and tools are available for assessing the evolutionary, structural and functional significance of the genes. The genes' and their products' structure, their structural changes and evolutionary significance can be assessed by a homology-based approach or simple information searching by taking advantage of the publically available genomic and protein databases (e.g., NCBI Genomic, NCBI conserved domains, NCBI GenBank Proteins, UniProtKB-Swiss-Prot, UniprotKB-PIR, and UniProtKB-TrEMBL, Pfam, Map Viewer). The genes can be functionally classified by using the Gene Ontologies (GO) and establishing pathway associations by using, for example, the Kyoto Encyclopedia of Genes and Genomes (KEGG). The function of the genes can be analyzed not only by an individual component but also by systematic approaches in the context of complex cellular processes (Hecker, Lambeck, Toepfer, van Someren, & Guthke, 2009). *In silico* pathway and network analyses – Ingenuity (http://www.ingenuity.com), Ariadne (http://www.ariadnegenomics.com/), Metacore (http://www.genego.com/metacore.php), PANTHER (Thomas et al., 2003), The Database for

Annotation, Visualization and Integrated Discovery (DAVID; Dennis et al., 2003) – can illustrate novel potential underlying neurobiological mechanisms and delineate neuronal circuitries that may translate genetic effects into behavior.

Due to the complexity of human brain function, it has been very difficult to generate animal models that mimic human cognition and cognitive disorders. The hypothesized role of novel genes in the underlying physiology of cognition or pathophysiology of disease involving cognition can be further tested with *in vitro* and *in vivo* experiments such as establishing aspects of gene function in cell culture experiments or transgenic animals. One of the examples of this top–down approach is the generation of a dominant-negative DISC1 transgenic mice, as DISC1 was found first in human studies in relation to schizophrenia, then targeted in the mouse, with measurable translatability as a model of schizophrenia (Hikida et al., 2007). This approach evidently can facilitate animal models relevant to cognition and potentially the development of novel therapeutics and gene-based therapies. Other approaches can consider the gene regulatory networks that the gene participates in, rather than the gene itself. These networks can be identified by computational biology and then these statistical models validated by animal experiments.

REPLICATION

The problem of failure to replicate a GWAS finding is common. Replication in the context of GWAS must be carefully considered. Based on current knowledge, the genetic bases of cognitive processes are not Mendelian; rather, multiple genetic influences can affect the same cognitive process (or an underlying mechanism such as signaling pathway). There are at least two levels at which replication of a study can be considered depending on the unit of analysis. At the most abstract level the replication can implicate the same pathway in a cognitive process or disorder. Replication at the molecular level, at a more conservative level, can include the identical sequence or SNP or genes depending on the research question. Most SNPs for GWASs are typically chosen as tagging SNPs (htSNPs), i.e., proxies for "blocks" of DNA (Balding, 2006). Tagging SNPs identify loci on a chromosomal region associated with the trait of interest. That region can harbor the "true" causal SNP which typically does not coincide with the htSNP, therefore the "true" SNP is not genotyped. Consequently, it is the block of DNA tagged by the htSNP(s) that becomes the focus for gene sequencing and subsequent studies of the causal molecular mechanisms.

The implications for replication of a GWAS are considerable. The most stringent replication requirement is finding the same allele of the original

SNP in the second study. However, this requirement can be misleading if the SNP to be replicated is a "tagging" SNP and the second population is not genetically homogeneous to the initial population (a likely event). In such cases, the same SNP may be implicated but the directionality may be different, or a different SNP that is located in the same chromosomal ("block") region may be implicated. This is more likely if the allelic frequency of the tagging SNP is not identical in the two populations. On the other hand, if the identified SNP in the first study is hypothesized to be a causal SNP, then failure to identify the same causal SNP in the second study, provided there is sufficient power (see Figures 1 and 2), is a failure of replication, a negative study.

As an example of the complexity of the replication, even in a "simple" association, we can consider the case of the phenotype of cystic fibrosis, which is known to be caused by an homozygous recessive genotype at the CFTR gene (cystic fibrosis transmembrane conductance regulator) (Zielenski, 2000). However, only 66% of all cases have a particular mutation, a 3-bp deletion called △F508, and more than a thousand mutations have been described in the remaining 33%. The frequency distribution of the mutations differs by ethnic groups. In fact, failure to find the particular △F508 mutation in cases with cystic fibrosis led to sequencing the entire gene, identifying the other mutations. The problems in replicating the non-△F508 mutations are obvious. The complexity of replication is likely to be much greater in non-Mendelian diseases or cognitive processes with multiple genetic influences.

Population stratification (PS) is also an important consideration in attempts to replicate a GWAS finding. Failure to recognize and correct for population stratifications is a major reason for failure in replicating genetic findings. PS occurs when allele frequencies differ between cases and controls due to ancestry differences, ethnic background or even "hidden" stratification, thus leading to spurious findings between a phenotype and unlinked candidate loci with either false positive or false negative results. To overcome this problem, different correction strategies have been proposed, however finding the most appropriate method to control for PS in not an easy task (Salvi et al., in press). Most of the current software used to analyze GWAS includes statistical controls and procedures to correct for PS, whenever present. With the most popular method, investigators calculate the inflation score due to PS with a statistic called *lambda* that represents the deviation from the median of a central chi-square distribution (Devlin, Bacanu, & Roeder, 2004), with PS considered to be present for values of lambda higher than 1. A method called Genomic Control (GC) (Bacanu, Devlin, & Roeder, 2000; Devlin & Roeder, 1999) is typically used to correct association tests for the estimated inflation score. Algorithms in the various software packages differ from each other in the way lambda is calculated as well as how GC is applied, leading to slightly, but sometimes consistently, different results (Salvi et al., in press). Investigators should be aware of the complexities related to PS, especially with the current trend in GWAS studies of considering large samples that rarely derive from a single population.

SUMMARY

Genes play a well documented role in cognition. Brain imaging obtained during performance of a cognitive task is a sensitive and quantitative reflection of cognition. Memory performance and regional brain activation during various cognitive tasks as measured in neuroimaging studies are examples of objective and quantitative measures that are usually heritable, and well suited as quantitative phenotypes (Blokland et al., 2008).

Brain imaging and genetic data can be integrated to enhance our understanding of cognition, and to discover unanticipated genetic influences on cognition. Using a summary measure of brain activation from a neuroimaging study as a cognitive phenotype in the context of a GWAS provides a strategy with increased statistical power over non-QT approaches. This QT approach can identify both known and unanticipated cognition-related genes.

There are several caveats related to findings with a QT GWAS approach. There is no consensus on the appropriate adjustments for multiple testing or how to calculate a GWAS threshold. The identified genes are merely candidates that require confirmation in an independent sample. The SNPs that identify these genes are primarily tagging SNPs, and therefore they do not necessarily convey any special meaning rather than "flagging" that particular gene or region. Study replication can be viewed at several levels and is dependent on the composition of the sample (population stratification). The choice of the QT in a GWAS is crucial to an interpretable result. The QT should represent an objective, quantitative and reliable measure that is relevant to the cognitive process. To the extent that the chosen trait is proximal to the genetic mechanism, the expected variance would be less, and increase the likelihood of identifying meaningful genes associated with the cognitive trait. The role of these newly identified genes require further exploration by *in silico* annotation, sequencing and molecular studies.

REFERENCES

Abdolmaleky, H. M., Faraone, S. V., Glatt, S. J., & Tsuang, M. T. (2004). Meta-analysis of association between the T102C polymorphism of the 5HT2a receptor gene and schizophrenia. *Schizophrenia Research*, *67*(1), 53–62.

Abi-Dargham, A., Mawlawi, O., Lombardo, I., Gil, R., Martinez, D., Huang, Y., et al. (2002). Prefrontal dopamine D1 receptors and working memory in schizophrenia. *Journal of Neuroscience, 22*(9), 3708–3719.

Almasy, L., Gur, R. C., Haack, K., Cole, S. A., Calkins, M. E., Peralta, J. M., et al. (2008). A genome screen for quantitative trait loci influencing schizophrenia and neurocognitive phenotypes. *American Journal of Psychiatry, 165*(9), 1185–1192.

Altshuler, D., Brooks, L. D., Chakravarti, A., Collins, F. S., Daly, M. J., & Donnelly, P. (2005). A haplotype map of the human genome. *Nature, 437*(7063), 1299–1320.

Anand, A., Li, Y., Wang, Y., Wu, J., Gao, S., Bukhari, L., et al. (2005). Antidepressant effect on connectivity of the mood-regulating circuit: An FMRI study. *Neuropsychopharmacology, 30*(7), 1334–1344.

Anderson, M. T., & Ter Braak, C. (2002). Permutation test for multi-factorial analysis of variance. *Journal of Statistical Computation and Simulation, 73*(2), 85–113.

Arking, D. E., Pfeufer, A., Post, W., Kao, W. H., Newton-Cheh, C., Ikeda, M., et al. (2006). A common genetic variant in the NOS1 regulator NOS1AP modulates cardiac repolarization. *Nature Genetics, 38*(6), 644–651.

Bacanu, S. A., Devlin, B., & Roeder, K. (2000). The power of genomic control. *American Journal of Human Genetics, 66*(6), 1933–1944.

Balding, D. J. (2006). A tutorial on statistical methods for population association studies. *Nature Reviews Genetics, 7*(10), 781–791.

Barabasi, A. L. (2007). Network medicine: From obesity to the "diseasome". *New England Journal of Medicine, 357*(4), 404–407.

Basile, V. S., Masellis, M., Potkin, S. G., & Kennedy, J. L. (2002). Pharmacogenomics in schizophrenia: The quest for individualized therapy. *Human Molecular Genetics, 11*(20), 2517–2530.

Benjamini, & Hochberg. (1995). Controlling the false discovery rate: A practical and powerful approach to multiple testing. *Journal of the Royal Statistical Society: Series B*, 57, 289–300.

Blokland, G. A., McMahon, K. L., Hoffman, J., Zhu, G., Meredith, M., Martin, N. G., et al. (2008). Quantifying the heritability of task-related brain activation and performance during the N-back working memory task: A twin fMRI study. *Biological Psychology, 79*(1), 70–79.

Botstein, D., & Risch, N. (2003). Discovering genotypes underlying human phenotypes: Past successes for mendelian disease, future approaches for complex disease. *Nature Genetics, 33*(Suppl.), 228–237.

Brzustowicz, L. M. (2008). NOS1AP in schizophrenia. *Current Psychiatry Reports, 10*(2), 158–163.

Calhoun, V. D., Liu, J., & Adali, T. (2009). A review of group ICA for fMRI data and ICA for joint inference of imaging, genetic, and ERP data. *Neuroimage, 45*(Suppl. 1), S163–S172.

Cardon, L. R., & Bell, J. I. (2001). Association study designs for complex diseases. *Nature Reviews Genetics, 2*(2), 91–99.

Cardon, L. R., & Palmer, L. J. (2003). Population stratification and spurious allelic association. *Lancet, 361*(9357), 598–604.

Chanock, S. J., Manolio, T., Boehnke, M., Boerwinkle, E., Hunter, D. J., Thomas, G., et al. (2007). Replicating genotype-phenotype associations. *Nature, 447*(7145), 655–660.

Chapman, J., & Clayton, D. (2007). Detecting association using epistatic information. *Genetic Epidemiology, 31*(8), 894–909.

Clayton, D., & McKeigue, P. M. (2001). Epidemiological methods for studying genes and environmental factors in complex diseases. *Lancet, 358*(9290), 1356–1360.

Corneveaux, J. J., Liang, W. S., Reiman, E. M., Webster, J. A., Myers, A. J., Zismann, V. L., et al. (in press). Evidence for an association between KIBRA and late-onset Alzheimer's disease. *Neurobiology of Aging*. Epub ahead of print, September 11, 2008.

de Bakker, P. I., Yelensky, R., Pe'er, I., Gabriel, S. B., Daly, M. J., & Altshuler, D. (2005). Efficiency and power in genetic association studies. *Nature Genetics, 37*(11), 1217–1223.

Dean, B., Keriakous, D., Scarr, E., & Thomas, E. A. (2007). Gene expression profiling in Brodmann's area 46 from subjects with schizophrenia. *Australian and New Zealand Journal of Psychiatry, 41*(4), 308–320.

Dennis Jr, G., Sherman, B. T., Hosack, D. A., Yang, J., Gao, W., Lane, H. C., et al. (2003). DAVID: Database for Annotation, Visualization, and Integrated Discovery. *Genome Biology, 4*(5), P3.

Devlin, B., & Roeder, K. (1999). Genomic control for association studies. *Biometrics, 55*(4), 997–1004.

Devlin, B., Bacanu, S. A., & Roeder, K. (2004). Genomic control to the extreme. *Nature Genetics*, 36(11), 1129–1130; author reply 1131.

Di Piero, V., Ferracuti, S., Sabatini, U., & Pantano, P. (1994). A cerebral blood flow study on tonic pain activation in man. *Pain, 56*(2), 167–173.

Dudbridge, F., & Gusnanto, A. (2008). Estimation of significance thresholds for genomewide association scans. *Genetics Epidemiology, 32*(3), 227–234.

Dudoit, S., & van der Laan, M. J. (2008). *Multiple testing procedures with application to genomics.* New York: Springer Verlag.

Durrant, C., Zondervan, K. T., Cardon, L. R., Hunt, S., Deloukas, P., & Morris, A. P. (2004). Linkage disequilibrium mapping via cladistic analysis of single-nucleotide polymorphism haplotypes. *American Journal of Human Genetics, 75*(1), 35–43.

Efron, B., & Tibshirani, R. (2002). Empirical bayes methods and false discovery rates for microarrays. *Genetic Epidemiology, 23*(1), 70–86.

Egan, M. F., Goldberg, T. E., Kolachana, B. S., Callicott, J. H., Mazzanti, C. M., Straub, R. E., et al. (2001). Effect of COMT Val108/158 Met genotype on frontal lobe function and risk for schizophrenia. *Proceedings of the National Academy of Sciences of the United States of America, 98*(12), 6917–6922.

Ennis, S., Jomary, C., Mullins, R., Cree, A., Chen, X., Macleod, A., et al. (2008). Association between the SERPING1 gene and age-related macular degeneration: A two-stage case-control study. *Lancet, 372*(9652), 1828–1834.

Evans, D. M., Marchini, J., Morris, A. P., & Cardon, L. R. (2006). Two-stage two-locus models in genome-wide association. *PLoS Genetics, 2*(9), e157.

Fallin, D., & Schork, N. J. (2000). Accuracy of haplotype frequency estimation for biallelic loci, via the expectation-maximization algorithm for unphased diploid genotype data. *American Journal of Human Genetics, 67*(4), 947–959.

Fallin, D., Cohen, A., Essioux, L., Chumakov, I., Blumenfeld, M., Cohen, D., et al. (2001). Genetic analysis of case/control data using estimated haplotype frequencies: Application to APOE locus variation and Alzheimer's disease. *Genome Research, 11*(1), 143–151.

Fisher, R. A. (1935). *Design of experiments.* Edinburgh: Oliver & Boyd.

Freedman, M. L., Reich, D., Penney, K. L., McDonald, G. J., Mignault, A. A., Patterson, N., et al. (2004). Assessing the impact of population stratification on genetic association studies. *Nature Genetics, 36*(4), 388–393.

Gianola, D., & de los Campos, G. (2008). Inferring genetic values for quantitative traits non-parametrically. *Genetical Research, 90*(6), 525–540.

Glazier, A. M., Nadeau, J. H., & Aitman, T. J. (2002). Finding genes that underlie complex traits. *Science, 298*(5602), 2345–2349.

Gottesman, II., & Gould, T. D. (2003). The endophenotype concept in psychiatry: Etymology and strategic intentions. *American Journal of Psychiatry, 160*(4), 636–645.

Gray, J. A., & Roth, B. L. (2007). Molecular targets for treating cognitive dysfunction in schizophrenia. *Schizophrenia Bulletin, 33*(5), 1100–1119.

Hecker, M., Lambeck, S., Toepfer, S., van Someren, E., & Guthke, R. (2009). Gene regulatory network inference: Data integration in dynamic models – A review. *Biosystems, 96*(1), 86–103.

Helgadottir, A., Thorleifsson, G., Manolescu, A., Gretarsdottir, S., Blondal, T., Jonasdottir, A., et al. (2007). A common variant on chromosome 9p21 affects the risk of myocardial infarction. *Science, 316*(5830), 1491–1493.

Hidalgo, C. A., Blumm, N., Barabasi, A. L., & Christakis, N. A. (2009). A dynamic network approach for the study of human phenotypes. *PLoS Computational Biology, 5*(4), e1000353.

Hikida, T., Jaaro-Peled, H., Seshadri, S., Oishi, K., Hookway, C., Kong, S., et al. (2007). Dominant-negative DISC1 transgenic mice display schizophrenia-associated phenotypes detected by measures translatable to humans. *Proceedings of the National Academy of Sciences of the United States of America, 104*(36), 14501–14506.

Hill, W. G., Goddard, M. E., & Visscher, P. M. (2008). Data and theory point to mainly additive genetic variance for complex traits. *PLoS Genetics, 4*(2), e1000008.

Hirschhorn, J. N., & Daly, M. J. (2005). Genome-wide association studies for common diseases and complex traits. *Nature Reviews Genetics, 6*(2), 95–108.

Hoffmann, T. J., Lange, C., Vansteelandt, S., & Laird, N. M. (2009). Gene-environment interaction tests for dichotomous traits in trios and sibships. *Genetic Epidemiology.*

Jiang, R., Tang, W., Wu, X., & Fu, W. (2009). A random forest approach to the detection of epistatic interactions in case-control studies. *BMC Bioinformatics, 10*(Suppl. 1), S65.

Jung, B. C., Jhun, M., & Song, S. H. (2006). A new random permutation test in ANOVA models. *Statistical Paper Springer-Verlag , 48, 2006*, 47–62.

Kamatani, N., Sekine, A., Kitamoto, T., Iida, A., Saito, S., Kogame, A., et al. (2004). Large-scale single-nucleotide polymorphism (SNP) and haplotype analyses, using dense SNP Maps, of 199 drug-related genes in 752 subjects: The analysis of the association between uncommon SNPs within haplotype blocks and the haplotypes constructed with haplotype-tagging SNPs. *American Journal of Human Genetics, 75*(2), 190–203.

Karlsgodt, K. H., Glahn, D. C., van Erp, T. G., Therman, S., Huttunen, M., Manninen, M., et al. (2007). The relationship between performance and fMRI signal during working memory in patients with schizophrenia, unaffected co-twins, and control subjects. *Schizophrenia Research, 89*(1–3), 191–197.

Keedwell, P., Drapier, D., Surguladze, S., Giampietro, V., Brammer, M., & Phillips, M. (in press). Neural markers of symptomatic improvement during antidepressant therapy in severe depression: Subgenual cingulate and visual cortical responses to sad, but not happy, facial stimuli are correlated with changes in symptom score. *Journal of Psychopharmacology.* Epub ahead of print, September 18, 2008.

Kennedy, J. L., Farrer, L. A., Andreasen, N. C., Mayeux, R., & St George-Hyslop, P. (2003). The genetics of adult-onset neuropsychiatric disease: Complexities and conundra? *Science, 302*(5646), 822–826.

Kruglyak, L. (2008). The road to genome-wide association studies. *Nature Reviews Genetics, 9*(4), 314–318.

Lander, E., & Kruglyak, L. (1995). Genetic dissection of complex traits: Guidelines for interpreting and reporting linkage results. *Nature Genetics, 11*(3), 241–247.

Lee, S. H., van der Werf, J. H., Hayes, B. J., Goddard, M. E., & Visscher, P. M. (2008). Predicting unobserved phenotypes for complex traits from whole-genome SNP data. *PLoS Genetics, 4*(10), e1000231.

Lee, S. S., Sun, L., Kustra, R., & Bull, S. B. (2008). EM-random forest and new measures of variable importance for multi-locus quantitative trait linkage analysis. *Bioinformatics, 24*(14), 1603–1610.

Lewis, D. A., & Gonzalez-Burgos, G. (2008). Neuroplasticity of neocortical circuits in schizophrenia. *Neuropsychopharmacology, 33*(1), 141–165.

Lewontin, R. C. (1958). A general method for investigating the equilibrium of gene frequency in a population. *Genetics, 43*(3), 419–434.

Li, J. (2007). Marker selection for whole-genome association studies with two-stage designs using dense single-nucleotide polymorphisms. *BMC Proceedings*, *1*(Suppl. 1), S136.

Lin, S., Chakravarti, A., & Cutler, D. J. (2004). Exhaustive allelic transmission disequilibrium tests as a new approach to genome-wide association studies. *Nature Genetics*, *36*(11), 1181–1188.

Lohmueller, K. E., Pearce, C. L., Pike, M., Lander, E. S., & Hirschhorn, J. N. (2003). Meta-analysis of genetic association studies supports a contribution of common variants to susceptibility to common disease. *Nature Genetics*, *33*(2), 177–182.

Manly (1997). *Randomization, bootstrap, and monte carlo methods in biology* (2nd ed.). London: Chapman and Hall.

Manoach, D. S., Press, D. Z., Thangaraj, V., Searl, M. M., Goff, D. C., Halpern, E., et al. (1999). Schizophrenic subjects activate dorsolateral prefrontal cortex during a working memory task, as measured by fMRI. *Biological Psychiatry*, *45*(9), 1128–1137.

McCarthy, M. I., Abecasis, G. R., Cardon, L. R., Goldstein, D. B., Little, J., Ioannidis, J. P., et al. (2008). Genome-wide association studies for complex traits: Consensus, uncertainty and challenges. *Nature Reviews Genetics*, *9*(5), 356–369.

McClearn, G. E., Johansson, B., Berg, S., Pedersen, N. L., Ahern, F., Petrill, S. A., et al. (1997). Substantial genetic influence on cognitive abilities in twins 80 or more years old. *Science*, *276*(5318), 1560–1563.

Menon, V., Anagnoson, R. T., Mathalon, D. H., Glover, G. H., & Pfefferbaum, A. (2001). Functional neuroanatomy of auditory working memory in schizophrenia: Relation to positive and negative symptoms. *Neuroimage*, *13*(3), 433–446.

Meyer-Lindenberg, A., & Weinberger, D. R. (2006). Intermediate phenotypes and genetic mechanisms of psychiatric disorders. *Nature Reviews Neuroscience*, *7*(10), 818–827.

Meyer-Lindenberg, A., Nichols, T., Callicott, J. H., Ding, J., Kolachana, B., Buckholtz, J., et al. (2006). Impact of complex genetic variation in COMT on human brain function. *Molecular Psychiatry*, 11(9), 867–877, 797.

Moore, J. H. (2003). The ubiquitous nature of epistasis in determining susceptibility to common human diseases. *Human Heredity*, *56*(1–3), 73–82.

Moore, J. H. (2008). Analysis of gene-gene interactions. *Current Protocols in Human Genetics, Chapter 1*, Unit 1.14.

Papassotiropoulos, A., Stephan, D. A., Huentelman, M. J., Hoerndli, F. J., Craig, D. W., Pearson, J. V., et al. (2006). Common Kibra alleles are associated with human memory performance. *Science*, *314*(5798), 475–478.

Perneger, T. V. (1998). What's wrong with Bonferroni adjustments. *British Medical Journal*, *316*(7139), 1236–1238.

Phillips, P. C. (2008). Epistasis: The essential role of gene interactions in the structure and evolution of genetic systems. *Nature Reviews Genetics*, *9*(11), 855–867.

Potkin, Kennedy, & Basile. (2002a). Brain imaging and pharmacogenetics in Alzheimer's disease and schizophrenia. In B. Lerer (Ed.), *Pharmacogenetics of psychotropic drugs.* Cambridge: Cambridge University Press.

Potkin, Kennedy, & Basile (2002b). Combining brain imaging and pharmacogenetics in understanding clinical response in Alzheimer's disease and schizophrenia. In B. Lerer (Ed.), *Pharmacogenetics of psychotropic drugs* (pp. 391–400). Cambridge: Cambridge University Press.

Potkin, Anand, R., Alphs, L., & Fleming, K. (2003). Neurocognitive performance does not correlate with suicidality in schizophrenic and schizoaffective patients at risk for suicide. *Schizophrenia Research*, 59(1), 59–66.

Potkin, S. G., Turner, J. A., Brown, G. G., McCarthy, G., Greve, D. N., Glover, G. H., et al. (2009a). Working memory and DLPFC inefficiency in schizophrenia: The FBIRN study. *Schizophrenia Bulletin*, *35*(1), 19–31.

Potkin, S. G., Turner, J. A., Fallon, J. A., Lakatos, A., Keator, D. B., Guffanti, G., et al. (2009b). Gene discovery through imaging genetics: Identification of two novel genes associated with schizophrenia. *Molecular Psychiatry, 14*(4), 416–428.

Potkin, S. G., Turner, J. A., Guffanti, G., Lakatos, A., Fallon, J. H., Nguyen, D. D., et al. (2009c). A genome-wide association study of schizophrenia using brain activation as a quantitative phenotype. *Schizophrenia Bulletin, 35*(1), 96–108.

Price, A., Patterson, N., Plenge, R., Weinblatt, M., Shadick, N., & Da, R. (2006). Principal components analysis corrects for stratification in genome-wide association studies. *Nature Genetics, 38*(8), 904–909.

Purcell, S., Cherny, S. S., & Sham, P. C. (2003). Genetic power calculator: Design of linkage and association genetic mapping studies of complex traits. *Bioinformatics, 19*(1), 149–150.

Purcell, S., Neale, B., Todd-Brown, K., Thomas, L., Ferreira, M. A., Bender, D., et al. (2007). PLINK: A tool set for whole-genome association and population-based linkage analyses. *American Journal of Human Genetics, 81*(3), 559–575.

Robbins, T. W., & Roberts, A. C. (2007). Differential regulation of fronto-executive function by the monoamines and acetylcholine. *Cerebral Cortex, 17*(Suppl. 1), i151–i160.

Rothman, K. J. (1990). Statistics in nonrandomized studies. *Epidemiology, 1*(6), 417–418.

Rothman, K. J. (1998). *Modern epidemiology* (2nd ed.). Philadelphia: Lippincott Williams & Wilkins.

Roffman, J. L., Weiss, A. P., Goff, D. C., Rauch, S. L., & Weinberger, D. R. (2006). Neuroimaging-genetic paradigms: A new approach to investigate the pathophysiology and treatment of cognitive deficits in schizophrenia. *Harvard Review of Psychiatry, 14*(2), 78–91.

Salvi, E., Guffanti, G., Orro, A., Lupoli, S., Torri, F., Potkin, S., et al. (in press). Ancestry correction in genome-wide association studies: Comparison of different methods to control for population stratification. *Briefings in Bioinformatics.*

Samani, N. J., Erdmann, J., Hall, A. S., Hengstenberg, C., Mangino, M., Mayer, B., et al. (2007). Genomewide association analysis of coronary artery disease. *New England Journal of Medicine, 357*(5), 443–453.

Seasholtz, T. M., Wessel, J., Rao, F., Rana, B. K., Khandrika, S., Kennedy, B. P., et al. (2006). Rho kinase polymorphism influences blood pressure and systemic vascular resistance in human twins: Role of Heredity. *Hypertension, 47*(5), 937–947.

Serre, D., Montpetit, A., Pare, G., Engert, J. C., Yusuf, S., Keavney, B., et al. (2008). Correction of population stratification in large multi-ethnic association studies. *PLoS ONE, 3*(1), e1382.

Shenton, M. E., Dickey, C. C., Frumin, M., & McCarley, R. W. (2001). A review of MRI findings in schizophrenia. *Schizophrenia Research, 49*(1–2), 1–52.

Shifman, S., Bronstein, M., Sternfeld, M., Pisante-Shalom, A., Lev-Lehman, E., Weizman, A., et al. (2002). A highly significant association between a COMT haplotype and schizophrenia. *American Journal of Human Genetics, 71*(6), 1296–1302.

Skol, A. D., Scott, L. J., Abecasis, G. R., & Boehnke, M. (2007). Optimal designs for two-stage genome-wide association studies. *Genetic Epidemiology, 31*(7), 776–788.

Slawski, M., Daumer, M., & Boulesteix, A. L. (2008). CMA: A comprehensive Bioconductor package for supervised classification with high dimensional data. *BMC Bioinformatics, 9*, 439.

Stranger, B. E., Forrest, M. S., Clark, A. G., Minichiello, M. J., Deutsch, S., Lyle, R., et al. (2005). Genome-wide associations of gene expression variation in humans. *PLoS Genetics, 1*(6), e78.

Tamminga, C. A. (2006). The neurobiology of cognition in schizophrenia. *Journal of Clinical Psychiatry, 67*(Suppl. 9), 9–13; discussion 36–42.

Teo, Y. Y. (2008). Common statistical issues in genome-wide association studies: A review on power, data quality control, genotype calling and population structure. *Current Opinion in Lipidology, 19*(2), 133–143.

Thomas, D. C. (2006). Are we ready for genome-wide association studies? *Cancer Epidemiology Biomarkers Prevention, 15*(4), 595–598.

Thomas, P. D., Campbell, M. J., Kejariwal, A., Mi, H., Karlak, B., Daverman, R., et al. (2003). PANTHER: A library of protein families and subfamilies indexed by function. *Genome Research*, *13*(9), 2129–2141.

Thomas, D. C., Haile, R. W., & Duggan, D. (2005). Recent developments in genomewide association scans: A workshop summary and review. *American Journal of Human Genetics*, *77*(3), 337–345.

Vawter, M. P., Shannon Weickert, C., Ferran, E., Matsumoto, M., Overman, K., Hyde, T. M., et al. (2004). Gene expression of metabolic enzymes and a protease inhibitor in the prefrontal cortex are decreased in schizophrenia. *Neurochemical Research*, *29*(6), 1245–1255.

Walters, J. T., & Owen, M. J. (2007). Endophenotypes in psychiatric genetics. *Molecular Psychiatry*, *12*(10), 886–890.

Wang, L. Y., Comaniciu, D., & Fasulo, D. (2006). Exploiting interactions among polymorphisms contributing to complex disease traits with boosted generative modeling. *Journal of Computational Biology*, *13*(10), 1673–1684.

Weinberger, D. R., Berman, K. F., & Zec, R. F. (1986). Physiologic dysfunction of dorsolateral prefrontal cortex in schizophrenia: I. Regional cerebral blood flow evidence. *Archives of General Psychiatry*, *43*(2), 114–124.

Westfall, P. H., & Young, S. S. (1993). *Resampling-based multiple testing: Examples and methods for p-value adjustment*. New York: John Wiley & Sons.

WTCCC (2007). Genome-wide association study of 14,000 cases of seven common diseases and 3,000 shared controls. *Nature*, *447*(7145), 661–678.

Yamada, R., & Okada, Y. (2009). An optimal dose-effect mode trend test for SNP genotype tables. *Genetic Epidemiology*, *33*(2), 114–127.

Ziegler, A., Konig, I. R., & Thompson, J. R. (2008). Biostatistical aspects of genome-wide association studies. *Biometrical Journal*, *50*(1), 8–28.

Zielenski, J. (2000). Genotype and phenotype in cystic fibrosis. *Respiration*, *67*(2), 117–133.

Zondervan, K. T., & Cardon, L. R. (2004). The complex interplay among factors that influence allelic association. *Nature Reviews Genetics*, *5*(2), 89–100.

Zondervan, K. T., & Cardon, L. R. (2007). Designing candidate gene and genome-wide case-control association studies. *Nature Protocols*, *2*(10), 2492–2501.

COGNITIVE NEUROPSYCHIATRY
2009, 14 (4/5), 419–450

Psychology Press
Taylor & Francis Group

Cognitive ontologies for neuropsychiatric phenomics research

Robert M. Bilder[1,3], Fred W. Sabb[1], D. Stott Parker[2], Donald Kalar[3], Wesley W. Chu[2], Jared Fox[2], Nelson B. Freimer[1], and Russell A. Poldrack[1,3]

[1]*Jane & Terry Semel Institute for Neuroscience & Human Behavior at UCLA, Los Angeles, CA, USA*, [2]*Department of Computer Science, Henry Samueli School of Engineering at UCLA, Los Angeles, CA, USA, and* [3]*Department of Psychology at UCLA, Los Angeles, CA, USA*

Now that genome-wide association studies (GWAS) are dominating the landscape of genetic research on neuropsychiatric syndromes, investigators are being faced with complexity on an unprecedented scale. It is now clear that *phenomics*, the systematic study of phenotypes on a genome-wide scale, comprises a rate-limiting step on the road to genomic discovery. To gain traction on the myriad paths leading from genomic variation to syndromal manifestations, informatics strategies must be deployed to navigate increasingly broad domains of knowledge and help researchers find the most important signals. The success of the Gene Ontology project suggests the potential benefits of developing schemata to represent higher levels of phenotypic expression. Challenges in cognitive ontology development include the lack of formal definitions of key concepts and relations among entities, the inconsistent use of terminology across investigators and time, and the fact that relations among cognitive concepts are not likely to be well represented by simple hierarchical "tree" structures. Because cognitive concept labels are labile, there is a need to represent empirical findings at the cognitive test indicator level. This level of description has greater consistency, and benefits from operational definitions of its concepts and relations to quantitative data. Considering cognitive test indicators as the foundation of cognitive ontologies carries several implications, including the likely utility of cognitive task taxonomies. The concept of cognitive "test speciation" is introduced to mark the evolution of paradigms

Correspondence should be addressed to Robert M. Bilder, Jane & Terry Semel Institute for Neuroscience & Human Behavior, 740 Westwood Plaza, C8-849, Los Angeles, CA 90095, USA. E-mail: rbilder@mednet.ucla.edu

This work was supported by USPHS grants including the NIH Roadmap Initiative Consortium for Neuropsychiatric Phenomics, including linked awards UL1DE019580 (RMB, FWS), RL1MH083268 (NBF), RL1LM009833 (DSP, WWC, JF), and The Cognitive Atlas (R01MH082795; RAP, DK). The development of this work was also supported by the NIH Roadmap Exploratory Center for Interdisciplinary Research (P20RR020750), and the Tennenbaum Center for the Biology of Creativity.

http://www.psypress.com/cogneuropsychiatry DOI: 10.1080/13546800902787180

sufficiently unique that their results cannot be "mated" productively with others in meta-analysis. Several projects have been initiated to develop cognitive ontologies at the Consortium for Neuropsychiatric Phenomics (www.phenomics.ucla.edu), in the hope that these ultimately will enable more effective collaboration, and facilitate connections of information about cognitive phenotypes to other levels of biological knowledge. Several free web applications are available already to support examination and visualisation of cognitive concepts in the literature (PubGraph, PubAtlas, PubBrain) and to aid collaborative development of cognitive ontologies (Phenowiki and the Cognitive Atlas). It is hoped that these tools will help formalise inference about cognitive concepts in behavioural and neuroimaging studies, and facilitate discovery of the genetic bases of both healthy cognition and cognitive disorders.

Keywords: Phenomics; Endophenotypes; Genetics; Cognition; Informatics.

INTRODUCTION

We live in an exciting scientific era marked by daily discoveries about the biological bases of brain function, and are deluged with findings from functional imaging, large-scale genetic association studies, and other work of high relevance to cognitive neuropsychiatry from its many cognate disciplines. We suggest that the development of cognitive ontologies may be important to help manage and digest this knowledge, and that application of cognitive ontologies may yield unanticipated insights into brain–behaviour relations. We first highlight some sources of this complexity that have arisen from efforts to find associations between genotypic variation and complex neuropsychiatric phenotypes, which have stimulated the development of bioinformatics and "phenomics" as new disciplines. It is then suggested that cognitive phenotypes may provide a useful level of analysis to help bridge the gap between genome and syndrome. Subsequent sections aim to familiarise the reader with ontologies as frameworks for formalising knowledge about specific concepts and their interrelations within a domain, and consider the unique challenges of designing ontologies to represent cognitive concepts, cognitive tests, and their relations to other kinds of knowledge. Finally, we describe works in progress that are designed to advance cognitive ontology development, the modelling of cognitive concepts, and the analysis of cognitive task effects, leveraging collaborative networks that have been enabled by the world wide web.

MANAGING COMPLEXITY IN THE "HUMAN PHENOME PROJECT"

In the "good old days" of neuropsychiatric genetics research—2 years ago— the literature was bustling with reports suggesting association of "candidate genes" with a range of phenotypes. Some of the phenotypes were traditional diagnostic categories, while others were "endophenotypes", including cognitive, electrophysiological, or neuroimaging measures. Now that we have reports from the first generation of genome-wide association studies (GWAS), the landscape is markedly different. Together with some disappointing failures to replicate prior candidate gene studies, the GWAS have prompted reconsideration and cast scepticism on findings that are not significant at a genome-wide level, and to consider findings "significant" today, suggested probability levels range from $p < 10^{-5}$ to $p < 10^{-7}$ (for more specific guidelines, see Freimer & Sabatti, 2004, 2005). Some suggest that we currently possess no strong candidate genetic loci for neuropsychiatry research (Flint & Munafo, 2007). Meanwhile, recent GWAS outside neuropsychiatry are providing novel leads at a rate that strains human capacity for comprehension, and prompting critical reevaluation of basic strategies for discovery in biomedicine (Frayling, 2007).

Initial GWAS results already make two critical points for neuropsychiatry research: (1) If we stick with conventional diagnostic categories as phenotypes, we are going to need very large samples to detect very small effects; (2) even if we are successful in defining "endophenotypes" or intermediate phenotypes, it remains unclear that these will possess "simpler genetic architecture". Given the large number and scale of GWAS now targeting neuropsychiatric phenotypes, it seems inevitable that we will soon possess a large number of leads—both genetic and phenotypic—that will require follow-up. If there are hundreds, or perhaps thousands of genetic variants involved (not to mention their interactions with both other genetic variants and environmental effects), and myriad phenotypes to consider, how will we prioritise leads for mechanistic research that will inform theories of pathophysiology and the development of rational treatments?

Bioinformatics strategies are helping develop a "bottom-up" scaffold enabling researchers to make the initial steps upwards from the human genome to the human proteome, and already can help constrain hypotheses with knowledge of the biochemical signalling pathways that have been associated with key genomic variants. But the ultimate vision of connecting this basic biological knowledge to the "higher" and more complex phenotypes comprising neural systems functions, cognitive abilities, neuropsychiatric symptoms and syndromes, will require more systematic and formal descriptions of these entities and their relations to each other and their putative biological underpinnings. Much as the Gene Ontologies project has succeeded in advancing genomics research, we suggest that cognitive ontologies can help advance research in *neuropsychiatric phenomics*—the systematic study of neuropsychiatric phenotypes on a genome-wide scale.

New tools to manage complexity can help increase our odds of making bets with solid payoffs in biomedical discovery. We may soon be poised to overcome current obstacles in genotyping, including detection of copy number variations and rare variants; even epigenome-wide arrays may be available. But we are likely still to confront significant obstacles converting large lists of genetic variants into tractable biological research programmes that will identify mechanisms by which these genes exert their effects on complex systems and syndromal levels. Within the Consortium for Neuropsychiatric Phenomics at UCLA (www.phenomics.ucla.edu), we have used a simple schematic scaffold for translational neuropsychiatric research from genome to syndrome, using seven levels (see Figure 1). Seven levels were selected not because we believe this reflects accurately the vast terrain connecting genome to syndrome, but rather because humans generally have difficulty maintaining in mind or discerning a larger number of categories.

Even given this dramatic simplification, it is easy to see that relating genome to syndrome with rational mechanistic hypotheses is a vast task. Kendler (2005) graphically highlighted that genotype-to-phenotype relations are "many-to-many" (i.e., any given genotype influences multiple phenotypes (pleiotropy), and a given phenotype has multiple genotypic contributions). But these many-to-many relations exist *between each level* as we proceed from genome to syndrome (i.e., a single protein may affect multiple cellular systems, and many proteins are required in any one cellular system; a single kind of

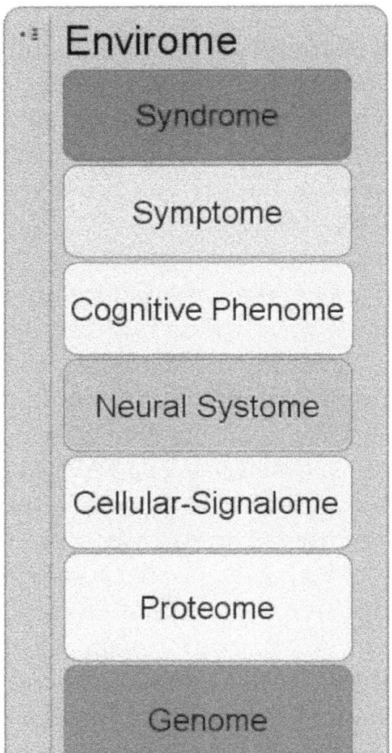

Figure 1. Simplified schematic of multilevelled "-omics" domains for cognitive neuropsychiatry.

cellular system has ramifications in multiple neural systems, and each neural system depends on manifold cellular systems; and so on). The immediate conclusion is that there can be exponential expansion of effects as one follows paths up the hierarchy (pleiotropic expansion) and that any given phenotype may be affected by very large numbers of genes (polygenic expansion).

Simple back-of-the-envelope calculations help put in perspective the scope of the challenges. Starting with the simplest assumption that there is a single gene with syndromal manifestations, and conservative estimates of pleiotropy at each level (i.e., that a single variant may affect p number of proteins, that each protein affects c number of cellular systems or signalling pathways, and so on), the influence of any single gene explodes through the higher phenomic levels. For example, a single gene with five-fold expansion across each of seven levels of phenomic expression will yield 15,625 effects. If the expansion were 10-fold at each level, there would be a million syndromal variants for each genetic variant.

This expansion logic also applies to top-down analyses of polygenic expansion. Imagine we seek genetic contributions to a single syndrome. Assuming the syndrome is defined by some number s of symptoms, and that each symptom has p cognitive underpinnings, and so forth, it is easy to appreciate that thousands of genes may contribute to complex phenotypes, and indeed that some brain-related phenotypes may be affected by substantial portions of the entire genome.

Similar conclusions are reached by computing the shared variance between levels in a seven-level hierarchy. For example, say we are interested in finding a gene that explains 25% of the variance in a complex syndromal phenotype. This demands that the average shared variance between levels is 80%, or in other words that each level must correlate approximately .9 with the next level. A more conservative but still optimistic estimate of 50% shared variance between levels (still demanding a correlation of .7, which is not far from the upward limits of reliability for psychiatric diagnosis, symptom rating scales, and cognitive phenotype measurements), yields shared variance between gene and syndrome of 1.6%. This might be considered the best case, possibly rational scenario. A more rational scenario is that in which approximately 20% of variance is shared across levels. This is more consistent so far with the typical correlations and effect sizes seen in psychiatry research, where correlations of cognitive measures to symptoms, or brain imaging parameters to cognitive dimensions, are not infrequently in the range of .4 to .5. In this scenario the shared variance of a genetic variant with a complex syndromal variant is .01%. A simple additive genetic model would thus require some 5000 genetic variants to explain a heritability of 50%, which is in the range of that observed for many cognitive, personality, or diagnostic phenotypes. The idea that this last scenario is more realistic than either of the former ones is supported by some reviews suggesting that genetic variation may explain only ~20% of variation at the level of the *transcript*, and that less than 2.5% of variance was shared with "higher" phenotypes, regardless of their putative proximity to the genetic level (Flint & Munafo, 2007). This line of reasoning calls into question the notion that gene discovery will be advanced substantially by the study of any particular "endophenotype", given that so far these have not often demonstrated a much more robust genetic signal, or "simpler genetic architecture".

Two basic strategies have been proposed to overcome these daunting challenges. Plan A, which we might label the "massively univariate" approach, is to increase sample sizes to detect the modest effects that appear most likely to characterise associations between genes and high-level phenotypes. It remains unclear how large these samples will need to be before we will find genetic variants associated with syndromal phenotypes like schizophrenia or bipolar disorder, but recent evidence suggests that such samples will likely exceed 10,000, and perhaps 50,000, before variants with robust genome-wide significance are observed. Two recent GWAS studies focusing on schizophrenia and bipolar disorder phenotypes, examining sample sizes exceeding 10,000 individuals, have reported a handful of strong targets (Ferreira et al., 2008; O'Donovan et al., 2008). Assuming that results from these studies stand up to replication, it is conspicuous that there is virtually no overlap with the long lists of candidate genes that have been suggested so far for these syndromes.

Plan B, which might be labelled the "multivariate" approach, or the "phenomics" approach, is to develop strategies to increase the magnitude of variance shared between genotype and phenotype through clever redefinition of genotype or phenotype, and the paths that relate genotype to phenotype. The cardinal premise of the "endophenotype" strategy is that there exist phenotypes "closer to the gene" that will share more variance with real gene effects. Although the Flint and Munafo (2007) survey does not offer much cause for optimism, it should be recognised that this strategy has seldom been employed in GWAS studies so far, and that the potential increases in power of this strategy remain largely unexplored. Even when multiple phenotypes have been examined, these are often derived from a single level of analysis (e.g., multiple partially overlapping diagnostic schemata) that is far removed from putative biological substrates.

Although not yet widely exploited for GWAS studies, some work with candidate genes has suggested that effect sizes may be considerably larger when examining neural system phenotypes (e.g., functional MRI measurements) relative to diagnostic or behavioural phenotypes (Egan et al., 2001; Hariri et al., 2003). It is also possible that more specific neurocognitive phenotypes may have strong relations with individual gene effects, by virtue of being more closely related to the physiological processes actually impacted by the genes. For example, traditional measures of "executive function" such as the Wisconsin Card Sorting Test perseverative error score shared less than 5% variance with the COMT val^{158}met polymorphism, while more specific measures of cognitive set-shifting shared up to 40% variance with genotype (Bilder et al., 2002; Bilder, Volavka, Lachman, & Grace, 2004; Nolan, Bilder, Lachman, & Volavka, 2004). Such high estimates might be inflated by chance given application of the candidate gene approach with small sample sizes, but highlight the possibility that refined phenotyping may yield greater promise not only by virtue of increasing statistical power, but further by enhancing insight into plausible mechanisms.

Even these theoretically more refined phenotypes are still derived from a single level of analysis, and it remains unclear what advantages might be obtained by defining new phenotypes that span different levels of investigation. For example, rather than identifying a new and improved neuroimaging or cognitive phenotype, we might find both more power and great mechanistic insight from combining a historical phenotype, an imaging phenotype, a cognitive phenotype, and a symptom phenotype. For example, perhaps a stronger genetic association might be found for individuals with poor premorbid social function, grey matter volume reduction, poor working memory, and negative symptoms, than could be found for any one of these alone. This may seem a counterintuitive strategy from the experimental perspective, but certainly has parallels in other disease areas. For example, the diagnosis of cardiac valve dysfunction benefits greatly from combining history (e.g., of early rheumatic fever), with laboratory results (e.g., from EKG), and behavioural symptoms (e.g., shortness of breath on exertion).

The definition of novel multilevel, multivariate phenotypes may also benefit from advances in complexity theory, and confer substantially greater traction on what might seem to be insuperable obstacles. Particularly given the number of emergent properties putatively implicated in the traversal of biological levels from the genome to syndrome, our ability to identify convergences and self-organising principles may help constrain the explosive expansion of possibilities to more manageable subsets. Thus, rather than attempting to explain the high heritability (perhaps 80%) of a syndrome such as schizophrenia by the additive effects of some 800 genetic variations each independently contributing 0.1% to phenotypic variance, there is hope that a smaller number of genetic variants might be identified that interact and converge on a more modest number of critical biological pathways. Stuart Kauffman (1995) has written eloquently about the "order for free" that characterises large-scale networks, and applied these principles to problems as diverse as the origins of life and economics. Similar methods have been used to help understand self-organisation in neural networks specifically, and biological networks more generally (Sporns, Tononi, & Kotter, 2005; Tononi, Sporns, & Edelman, 1994; Sporns, & Edelman, 1999). Progress is already being made identifying the redundancy and degeneracy in gene networks and other biological systems using techniques that integrate information sciences and systems biology (Centler, Kaleta, di Fenizio, & Dittrich, 2008; Sridhar et al., 2007; Zheng, Yeh, Yang, Jang, & Chu, 2007). Application of similar methods to multilevel modelling of genome-to-syndrome hypotheses comprises a reasonable if not simple extension of these theoretical lines. Regardless of the specific methods that will be used, it is clear that novel strategies to effectively manage complexity of large-scale networks spanning different levels of investigation will be critical to advance the emerging discipline of phenomics.

WHY PICK THE COGNITIVE PHENOME AS A FOCAL POINT?

If one accepts the basic principle that there exists some roughly hierarchical set of causal mechanistic links that lead from the genome to the syndrome, it is rational to ask: Where to start? The strategy of conducting "phenome-wide" scanning (measuring all possible phenotypic variations from mRNA through complex behavioural syndromes) is obviously impossible in practical terms, and arguably impossible even in theory. Given imperfections in our knowledge, we may never realise what truly comprises the human phenome; in contrast, a complete sequence of an individual genome is straightforward, if currently a bit costly. Scientists must therefore choose a level of phenotypic complexity appropriate to their interest, to narrow the scope of possible investigation.

Some investigative teams are logically focusing on the human proteome, an attractive target given its "proximity" to the genome. But this does not offer clear traction for neuropsychiatry, given that so far few single protein anomalies have been identified as the culprits underlying complex psychiatric disease. To put this in perspective, consider the long and so-far frustrating search for an abnormal dopamine receptor protein in schizophrenia. Perhaps the recent GWAS findings implicating the L-type voltage gated calcium channel in bipolar disorder will prove to be more fruitful (Ferreira et al., 2008). Even if we do identify some critical proteins to help focus research, there will clearly remain substantial intervening biology to connect these protein variations via mechanistic hypotheses to neuropsychiatric symptoms and syndromes (Bilder, 2008).

In efforts to identify a rational starting point in the seven-layered schema, the CNP has suggested that the level of cognitive phenotype is attractive because: (a) the cognitive phenotypes may be rationally related to higher level symptoms and the syndromes that are defined by these; (b) cognitive phenotypes might offer links to neural systems level analysis, particularly given advances in functional neuroimaging that enable experimental manipulation of cognitive parameters and use of brain activation as the dependent variable; (c) at least some cognitive phenotypes appear to have homologues, or at least analogues, in other species, enabling links to a broader class of basic science studies.

Compelling arguments can be advanced for organising research programmes around alternate phenotypic nodes at other levels. For example, in the study of Alzheimer's disease, the diagnostic entity in the DSM-IV (American Psychiatric Association, 1994) for which there is the strongest consensus about pathophysiology, it might be considered rational to focus on amyloid deposition or tau-opathies, as organising phenotypes. Even in this example, however, it should be recognised that prior definitions of Alzheimer's disease may require revision, and that an extended range of related disease entities may emerge from this redefinition (Chiang, Lam, & Luo, 2008; Saura et al., 2005). For disorders that are less understood, including psychotic disorders, mood disorders, anxiety disorders, and most disorders first appearing in childhood and adolescence, a focus on cognitive phenotypes may provide a helpful link between the high level syndromes identified as public health concerns, and brain biology. In order to connect cognitive phenotypes to the syndromal level, it is further explicit in this formulation that phenomics will benefit from dimensional representation of psychiatric problems at the symptom level, and this practice is supported by the advantages of dimensional over discrete categorical approaches to the characterisation of psychiatric syndromes (Haslam, 2003; Haslam & Kim, 2002; Kraemer, 2007).

Narrowing the problem scope by selecting a phenotypic node as a starting point leaves us still with enormous complexity. To gain traction on the complex paths leading from genomic variation to syndromal manifestations, informatics strategies must be deployed to consolidate broader domains of knowledge and help researchers find the most important signals. Bioinformatics has grown enormously over the last decade, and generated repositories of biological knowledge at the genomic, transcript, proteomic, and even metabolomic levels. But higher phenomic levels are not yet well represented. Efforts so far to database higher level phenotypic knowledge (e.g., fMRI findings) have so far suffered, probably for multiple reasons, including concerns about the mechanisms and safeguards for data (and credit) sharing, and due to the complexity of the data themselves. But at least one substantive obstacle in these efforts has been the lack of a coherent conceptual system to organise these effects. Projects such as the Function Biomedical Informatics Network (FBIRN) have worked to resolve some of the problems with variability in data acquisition and data formatting parameters. It has proven more difficult to arrive at consensus on the optimal methods for describing the cognitive manipulations used as independent variables in these experiments. This poses a challenge for effective data sharing and synthesis, and further may make it difficult to generate appropriate inferences from neuroimaging data (Poldrack, 2006). A coherent consensus cognitive ontology could help such efforts realise their fullest potential.

WHAT ARE ONTOLOGIES AND WHY MIGHT THEY HELP?

Can ontologies help to reduce the complexity of the enormous problem space we are confronting? Despite ontology development seeing enormous growth over the last decade, many investigators in cognitive neuropsychiatry are not very familiar with ontology development or how ontologies can be used. The term ontology may be most familiar given its long history of usage in philosophy to mean "the study of the nature of being, existence, or reality in general and of its basic categories and their relations" (Wikipedia, accessed 4 October 2008). The term ontology has a different usage in computer and information sciences, as "a formal representation of a set of concepts within a domain and the relations between those concepts. It is used to reason about the properties of that domain, and may be used to define the domain" (Wikipedia, accessed 4 October 2008). A more concise definition of the term was offered by Gruber (2003, p. 199): "an ontology is a specification of a conceptualization". Gruber further emphasises the purpose of such explicit specification is to enable "knowledge sharing and reuse". It is precisely this capacity for developing shared representations of knowledge regarding cognitive phenotypes that we believe is of paramount importance for advancing the discipline so that emerging knowledge can be more readily shared within the cognitive neuroscience community, and so that links can be forged to other levels of both clinical and biological knowledge.

There are various degrees of formalisation possible in ontology development; at least four levels of progressively increased elaboration in ontology definition have been identified (Grethe, Gupta, Ludascher, & Martone, 2003):

- *Controlled vocabularies*: The first stage in most ontology development is the development of a lexicon, or set of terms representing the concepts in a specific domain. The vocabulary is "controlled" by establishing operational definitions of the terms, identification of synonyms, and identification of other properties of the terms (for example, part of speech, pluralisation, and so on). Ideally this will lead to a set of "standard names" for key concepts in the domain, which can then be used to foster greater agreement in concept representation by the users.
- *Simple taxonomies*: Once a set of terms reflecting the conceptual entities of the domain is identified, the next level of refinement is often the classification of these entities into a taxonomy, sometimes referred to as

a "class hierarchy". These class hierarchies are often represented as *directed acyclic graphs*, which enforce a unidirectional "flow" from superordinate categories to subordinate category members, and use a specific type of relation sometimes described as "is a" relationships. This kind of class hierarchy works well in certain concept areas (i.e., phylogenies of plants and animals; for example, a cat is a mammal, a mammal is an animal, and so forth).

- *Graph-like ontologies*: Many conceptual domains are not well modelled by simple class hierarchies, and more complex types of relation are required to represent the structure of the domain. It is likely that the domain of cognition is a case in point. Thus, while there might be some consensus that "episodic memory" is a subclass of "memory", other concepts such as "working memory" might be considered a component of "memory" or a component of "executive function", emphasising different aspects of this concept. Such ontologies may possess many unique relation types ("has a", "contained in", "causes", "activates", etc.). In cognitive ontologies, there are several particularly useful relation types. For example, cognitive task indicators (the specific variables that summarise performance on aspects of the task) are considered to *measure* certain cognitive concepts. Conversely, cognitive concepts may be "measured by" task indicators. Such ontologies may be rich with different kinds of relations but lack more formal semantics.
- *Logic-based ontologies*: More formal logic-based ontologies permit inference of relations that are logical consequences of formal concept definitions that in turn arise from formalised syntax and semantics. The goal of such ontologies is to afford automated inference or reasoning within the domain of the ontology. There are efforts to establish conventions for logic-based languages that standardise syntax and semantics for information interchange, including: CLIF—Common Logic Interchange Format; CGIF—Conceptual Graph Interchange Format; and XCL—eXtended Common Logic Markup Language, based on XML (for more information, see www.obitko.com).

For the domain of "cognition", there does not exist yet any agreed-upon lexicon, and a lexicon is necessary for developing any more elaborate ontology, so this represents a rational starting point. It is also clear that the concepts in the domain of cognition are unlikely to be well expressed by simple taxonomic structure, although such representations may be useful intermediate steps to help clarify relations among key concepts (a few examples are provided later). Ultimately, we believe the most functional ontologies may be graph-like, offering a high degree of flexibility in ontological specification, and acknowledging that our understanding of this domain may long remain only quasistructured and the conceptual entities themselves subject to revision as we learn more about relations between brain and behaviour.

COGNITIVE ONTOLOGY NUTS AND BOLTS: HOW CAN WE ANCHOR ABSTRACT COGNITIVE CONCEPTS?

To address this question it may help to flesh out more concretely what elements comprise a simple ontology. At the base of ontologies are specific examples or *instances* of the things that exist in the domain. These can be concrete things like individual people, animals, cars, or bottles of wine; or the things may be abstract like words or numbers (i.e., "thing" in this context does not signify a concrete object, but instead signifies existence—a "thing" can be distinguished from nothing).[1] Ontologies usually also contain *classes* or categories of things, and these are often arranged hierarchically, as noted earlier for simple taxonomies. *Relations* are used in ontologies to formalise the ways in which instances or classes can relate to each other (the classic "is a" relation is widely used to denote class membership, e.g., a cat is a mammal). *Attributes*, sometimes referred to as properties or features, are qualifiers that may apply to the things in the ontology. These may include attributes important to identifying class membership (e.g., "has sweat glands" and "has hair" are attributes of animals that help identify them as mammals), or attributes that help distinguish instances within a class (e.g., "is red" may distinguish one set of cars from others of the same make and model, but other makes and models may share this attribute).

For cognitive ontologies, the instances are relatively abstract given that mental entities are not observable. For example, consider what comprises an "instance" of the concept "memory": Is it a subjectively experienced "memorandum"? Is it a process? Is it a neural system-level representation reflecting connection weights among a set of neurons linked in a network? We believe there is a need for ontologies to help organise and sharpen definitions for the abstract concepts that are used in cognition research (e.g., is "working memory" a kind of "memory", and what are the exact distinctions among the terms "short-term memory" and "working memory" or between "episodic memory" and "declarative memory"?). Given the imprecision, inconsistency, and temporal instability in usage of cognitive concept terms, we suggest it is important to anchor these concepts at the measurement level. If this suggestion is accepted, instances are the results of specific measurements made on individual animals (including humans).

---

[1] Since ontologies are created to enable the *classification* of instances, they may not actually *contain* the instances themselves; nevertheless, it is important to consider the actual "things" that the ontology aims to classify.

Thus, while capturing useful aspects of our shared conceptualisations of cognitive processes on the one hand, an ontology of cognitive concepts can be usefully rooted in more concrete and stable measurement methods that provide the experimental and observational basis for the abstractions used in our discipline.

Figure 2 illustrates a simple class hierarchy for a cognitive concept and its relation to a cognitive measurement. The figure suggests that verbal declarative memory is a kind of declarative memory, and that declarative memory is a kind of memory. The figure also suggests that verbal list learning procedures can offer measures of verbal declarative memory (and thus also of declarative memory, and memory by inheritance), but that this relation is mediated specifically via an indicator (number of words recalled) on a specific verbal list learning test. It may be true that all verbal list learning tests possess indicators of verbal declarative memory, but other measures derived from these tests may not measure verbal declarative memory (for example, indicators for semantic clustering or proactive interference, although derived from the same test, may measure different constructs). The importance of this distinction is still more apparent with an example where indicators map to more distinctive concepts. Figure 3 shows how two different indicators derived from the Stroop Test measure two divergent cognitive concepts (one is the tendency to inhibit prepotent responses via the "classic Stroop effect", which is itself a contrast effect

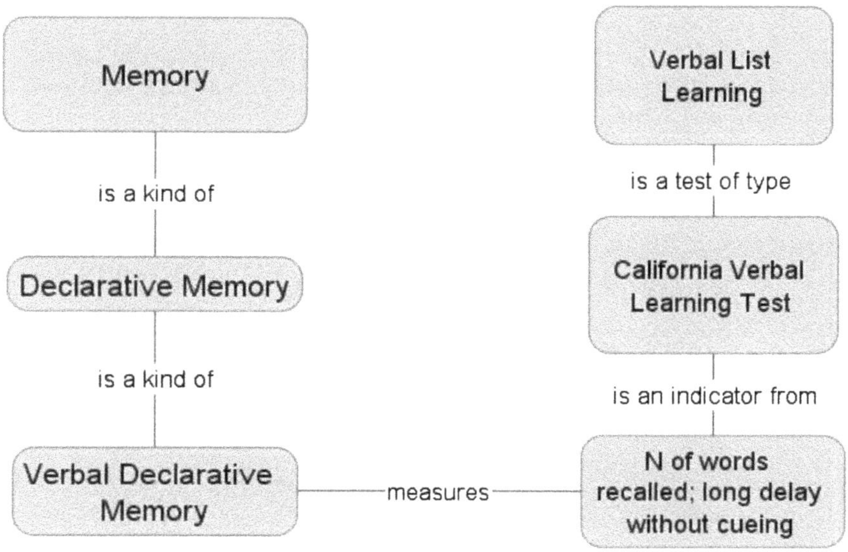

Figure 2. Alignment of cognitive concept and task hierarchies via measurement models.

between an "interference" condition and a control condition; the other is simple colour naming skill). Thus, it would be incorrect to say simply that the Stroop Test measures "response inhibition" without reference to the specific observable indicator that measures this function.

Figure 3 further highlights that some concepts (e.g., response inhibition) may be considered subclasses of multiple higher level concepts (e.g., is response inhibition better considered a kind of "cognitive control" or an "executive function"?). In prior work we have shown that there can be dramatic increases in usage of some concept labels (such as the term "cognitive control"), although there is considerably greater stability at the measurement level (for example, in the literature on heritability of cognitive control, there was not a single instance where the cognitive control construct was measured using an indicator that had not previously been considered an indicator of "working memory", "response inhibition", "task switching/set shifting", or "response selection") (Sabb et al., 2008). Anchoring cognitive concepts to observable indicators promotes greater consensus and interpretation at appropriate levels of abstraction, with greater immunity from variability in concept labelling.

Anchoring cognitive concepts to the measurement level further enables cognitive ontologies to serve as the organisational foundation for repositories of quantitative, empirical data that can in turn be used for automated meta-analysis and construction of multilevel causal models. For example, assume there are multiple indicators of a specific concept, and that we possess information about the covariance of those test indicators. Then the validity of the concept can be examined using the methods of covariance structure analysis (i.e., structural equation modelling, where the concept is defined as a latent construct, and the observed variables are the indicators of that construct). Structural equation modelling typically uses case-level data

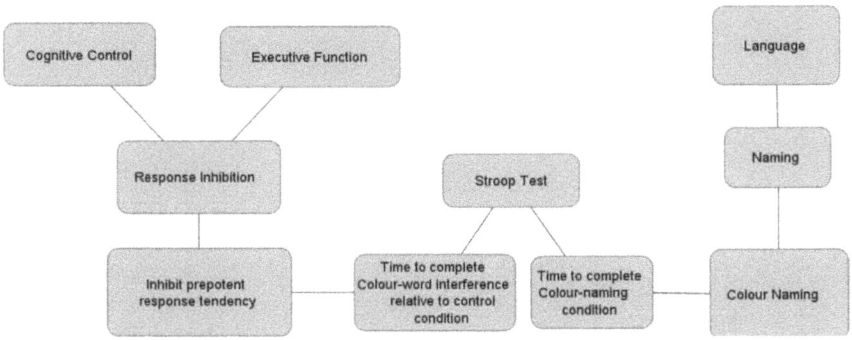

Figure 3. Alignment of cognitive concept and task hierarchies with multiple divergent indicators, and possible controversy over class membership within the concept hierarchy.

where each observation represents an individual subject and their scores on relevant indicators, but there is an emerging literature on Meta-Analytic Structural Equation Modelling, which enables analyses to be pooled at the study result level (Cheung, 2008; Furlow & Beretvas, 2005; Riley, Abrams, Sutton, Lambert, & Thompson, 2007). There are multiple challenges to assuring the validity of results derived from such analyses, centring on the degree to which results from different studies can reasonably be pooled. Given appropriate caution, however, there is promise that these tools will help foster our capacity to synthesise results from widely divergent studies and develop estimates of models spanning multiple empirical investigations.

It is logical to worry that anchoring cognitive concepts to the measurement level might reify concept definitions and constrain the specification of novel concepts. This is not a limitation, however, if the system for ontology specification is flexible and can be modified by individual investigators. For example, an investigator may have a completely novel theory and unique set of concept labels. This investigator could construct an individual ontology containing these new concepts and their interrelations, and also link these to existing empirical data, or provide the rationale as to why the existing empirical data are not relevant to these concepts, thus highlighting the need for new data to assess the validity of the novel theory. Strategies for coping with conflicting theoretical views are discussed further in a subsequent section. It should also be recognised that concept ontologies can be advanced *without* being anchored to any empirical data, and therefore it is entirely possible to pursue cognitive ontology development in a "top-down" fashion (i.e., from concept to task). On the other hand, there is a risk, in developing a "top-down" ontological model of a concept domain, that if it is not linked to empirical data, it will remain essentially unfalsifiable.

DO WE REALLY NEED COGNITIVE ONTOLOGIES?

It is reasonable to ask: Why bother with ontology development? There are multiple reasons for supporting development of cognitive ontologies to advance research on the genetics of, and treatments for psychiatric disorders.

- *Connections*: Links to other repositories of biological knowledge will be possible only if the knowledge regarding complex cognitive and neuropsychiatric phenotypes are rationally constructed and appropriately indexed. Current indices are too crude to enable meaningful inferences to be drawn. For example, the Medical Subject Headings (MeSH) that are the most widely used indexing system for medical literature contain only six terms under the concept "Cognition". These terms do not reflect very well current research in cognitive psychology and neuroscience (the complete list is: Awareness; Cognitive Dissonance; Comprehension; Consciousness; Imagination, which includes the subconcepts Dreams and Fantasy; and Intuition). Similarly, the components of ontologies in other repositories of biological knowledge are underdeveloped from the perspective of those interested in cognition (for example, searching for the term "cognition" in the National Library of Medicine's [NLM's] Entrez GENE database currently yields a total of only 83 hits and the same search in NLM's OMIM [Online Mendelian Inheritance in Man] yields only 54 hits [as of 8 October 2008]). These searches are clearly failing to show high sensitivity to a large literature already in existence. Good cognitive ontologies would enable literature reports and data to be linked far more effectively with relevant biological constructs and support both validation of existing hypotheses and exploration for development of novel hypotheses. As molecular libraries and proteomics knowledge bases continue to expand, cognitive ontologies may become critical links also in drug discovery by helping to narrow the range of potential targets and their possible mechanisms of action.

- *Formalisation*: The "structure" of cognitive and neuropsychiatric phenotypes remains too frequently the target of subjective speculation, and is strongly influenced by conceptual and semantic factors that are often independent of the underlying data. Perhaps in part due to the "fuzziness" of these conceptual distinctions, their imperfect relations to measurement models, and their loose validation with respect to underlying biological processes, it is easy for conclusions to be drawn that are inappropriately generalised, lead to inertia in certain research areas, along with replication failures and unproductive controversy. It is also important to the disciplines of psychology and psychiatry for the concepts we use to be clearly operationalised, to foster more effective public communication. A challenge to our disciplines is that many terms we use in special ways may overlap with lay language, and this can too often result in misinterpretation. Adoption of stricter definitions can help avoid misappropriation of our language.
- *Sharing*: Currently ideas are shared in publications, conferences, lab meetings, and on the Internet; these are the primary means for learning, communicating viewpoints, generating ideas, and designing hypotheses and experiments about cognition. But despite some widely read textbooks, there remains no collective repository of formalised knowledge about the structure of cognition, the methods used to measure its facets, or the relations of cognitive functions to behavioural disorders, brain function, or other biological characteristics. Given the demonstrated efficiency of collaborative networking applications like Wikipedia, we believe that is both technically feasible and desirable for a shared

knowledge base reflecting current thinking and argumentation about cognition and its relations to be developed.

If cognitive psychology and neuroscience were simple disciplines, independent of other areas of brain and biological sciences, there would be less need for ontology development. But cognitive psychology and neuroscience are intrinsically interdisciplinary sciences, and there is hope in coming years that substantial breakthroughs may emerge specifically at the interdisciplinary edges between cognitive sciences and other disciplines, particularly given the growth trajectories of genetics and neuroimaging. Cognitive ontologies will be important to maximise the potential for realising such breakthroughs.

CHALLENGES FOR COGNITIVE ONTOLOGY DEVELOPMENT: CONFLICT RESOLUTION AND LABILITY OF CONCEPT LABELS

Several of the challenges facing cognitive ontology development were mentioned above, including conspicuously the lack of formal definitions of key concepts and relations among entities, and the fact that cognitive concepts are not likely to be well represented by simple hierarchical "tree" structures. There are additional challenges because the usages of the concepts differ among investigators, and even the tasks used to measure cognitive constructs evolve over time. Some examples may clarify these points.

In Figure 3 a simple example was presented, in which the concept "response inhibition" was considered a kind of "cognitive control", but also a kind of "executive function". Resolving such a conflict might seem a simple matter of taste and of little consequence, but such choices ultimately can impact research directions, selection of methods, and approaches to synthesising results both within and across studies. It would be possible to define a class hierarchy and enforce unique parent–child relations, based on selected expert opinions. We believe that there is sufficient imprecision and ambiguity in the definition of cognitive concepts and their interrelations that such mixed parentage should be allowed, and open to argumentation. Thus, we are developing infrastructure to represent precisely such conflicts and enable active argumentation about the "correct" structure as perceived by the field, which we hope will be resolved by empirical evidence supporting one or the other point of view. Absent empirical substantiation, it is also possible to incorporate "voting" summaries, which can indicate which are majority and minority views, with annotation describing the rationale supporting each perspective. We hope that some approaches to classification will be pruned early. For example, the utility of a cognitive concept representing "frontal lobe functions" might be considered redundant, and could be obviated by creating relations of certain cognitive processes specifically to external validating measures of "frontal lobe function" as measured by fMRI activation or single unit activity in anatomically defined regions. Such tasks are simplified in part by reasonably well developed taxonomies of neuroanatomic structure (e.g., Foundational model of anatomy, Neuronames, and the modification of this utilised in the PubBrain application: www.PubBrain.org).

An interesting feature of cognitive concept labels is that these are not stable in use over time, and sometimes appear to reflect changes in fashion more than changes in either the underlying concept or the underlying measurements. For example, Jacobsen (1936) is usually credited with the initial observations of deficits in animals' ability to perform delayed response tasks following lesions to the dorsolateral prefrontal cortex. Subsequent studies revealed that this ability was associated with sustained activation in frontostriatal systems, and with unique firing patterns of pyramidal cells in the frontal lobe, during intervals in which animals were performing the delayed response tasks. Considerable work using this delayed response paradigm was conducted by Pribram and colleagues (for review see Pribram & McGuinness, 1975), Fuster and colleagues (e.g., Fuster & Alexander, 1971), Goldman-Rakic and colleagues (reviewed in Goldman-Rakic, 1987), and many others subsequently. Jacobsen referred to the delayed response behaviour as "immediate memory"; Pribram referred to it as "activation"; Fuster called it "short-term memory"; and Goldman-Rakic called it "working memory". Experiments and theory focusing on these processes also have used the terms "cognitive control" (Miller & Cohen, 2001; Posner & Snyder, 1975). Interestingly, "cognitive control" is considered in Wikipedia to be a synonym for "executive function", and within cognitive control some theorists have identified a key element as "context processing", which may include "maintenance" and "updating" components (Braver & Barch, 2002). Although each of these terms possess unique connotations, this example was used to emphasise that results of a single experiment (e.g., on a simple delayed response task) might be subject to a plethora of semantic labels, and is problematic precisely *because* these labels have distinct connotations.

The idea that popularity of concept labels may also affect usage *even when connotations are not different* is supported by a recent analysis that aimed to determine heritability for measures of "cognitive control" (Sabb et al., 2008). Since there was little genetic literature specifically mentioning "cognitive control", we enumerated a series of five terms most frequently cooccurring with "cognitive control" in the PubMed literature over the last decade (these were "working memory", "response inhibition", "response selection", and "task switching/set shifting"). We then sampled 30 papers that used each of these terms to determine which specific cognitive task measures were employed to assess each construct. The surprising result was, despite a clear ascendancy in use of the term "cognitive control" to describe results, there was not a single cognitive test that was uniquely associated with "cognitive control"; all of the specific measures had previously been used as an indicator of one of the other five concept labels. These findings further underscore the importance of anchoring cognitive concept labels at the measurement level.

TASK EVOLUTION AND TASK SPECIATION

Even if we are able to achieve higher consistency in labelling of results that are in fact consistent, there remains a challenge of identifying appropriate labels for tasks that have changed over time. Most tasks used in cognitive psychology research are not identical across different laboratories or even within the same laboratory over time. A major advantage of anchoring cognitive ontologies to the measurement level is that the strategy for determining changes in task properties is easier than tracking changes in concept definitions and usage. The process is easier because task parameters are usually (if not always) operationalised objectively, offering a clear basis to judge divergence in methods. The process is also easier because most tasks are based on prior tasks, and thus can more readily be considered descendants in a phylogenetic sense. The capacity to trace task modifications via their historical roots confers further advantages, by providing a phylogenetic perspective that may complement other methods for task classification. Thus, for example, rather than classifying tasks by the putative cognitive processes that these measure in a "top-down" manner, it may be possible in many cases to trace origins to some smaller set of "founder" tasks. By tracing the lineage of a specific task implementation reported in the literature, it is feasible to determine more readily exactly which parameters changed between versions, so that any differences in associated effects may be more rationally attributed to specific manipulations.

Although there are not yet many detailed historical surveys on the origins of cognitive tasks, when available these offer unique insights into the current "state of the art", and can offer perspective also on construct redefinitions that have emerged over time. For example, there is a very interesting analysis of the origins of the Wechsler IQ scales (Boake, 2002). We have constructed a diagram that summarises that history (available online at www.phenomics.ucla.edu); Figure 4 extracts only 12 of the 95 elements from that diagram that pertain specifically to the familiar Digit Span test (please see Boake, 2002, for complete citation information). Figure 4 illustrates how the digit span procedure has evolved over the last century, and further how concepts may have shifted regarding what functions are tapped by this test. According to Boake, Ebbinghaus considered this a measure of repeating sounds, while Galton and Jacobs felt this process was best labelled "prehension". Through earlier versions of the Wechsler Scales, factor analytic studies initially led digit span performance to be considered an index of "attention", and in studies of children a similar factor was labelled "freedom from distractibility". More recent factor analytic studies suggest digit span scores load most on a "working memory" factor (although this followed inclusion of another "working memory" test, the Letter–Number Sequencing test, into these analyses). The digits backward condition also may be the most widely cited example of the "manipulation" component of working memory, or "working with memory." The ephemeral quality of these labels belies the observation that key aspects of the actual test procedure have remained constant for 120 years.

If tasks can be classified based on their phylogenetic origins, it is then reasonable to ask when a task has evolved significantly from its predecessors, to the point that it might be considered a new task "species"? We believe this process can be operationalised by introducing the concept of cognitive task "speciation." Task speciation can be used to mark those events in test development that generate paradigms sufficiently unique that their results cannot be "mated" productively with others in meta-analysis. This determination remains a matter of some subjectivity, and rests with the individual investigator as to whether two sets of test results are sufficiently similar to be considered reasonable targets for meta-analytic combination. But by providing an objective framework, through a systematic analysis of specific task properties that have changed between versions, it may be superior to the methods used currently, which are usually not fully specified.

Taking the digit span procedures as an example (Figure 4), should we actually consider the WAIS-IV version to be the same task species as the original Ebbinghaus procedure? Multiple features of the task have changed, including the number of sequences, the rate of administration, the inclusion of the digits backwards condition, and of course the exact number that are to be repeated. Thus for many purposes one might argue that these are clearly different species, that should not be combined in meta-analyses. But what about results from the WAIS-III versus results from the WAIS-IV tasks with the same name? The WAIS-IV did modify the task to eliminate rhyming numbers; but did this really change the construct measurement significantly? For some investigators this difference might be trivial, but for someone interested in phonological processing and its effects, this difference might be of primary interest.

Despite such differences of opinion, we believe it is likely that clear taxonomies reflecting the origins of tasks and specifying the changes that have been introduced in different versions can be illuminating and of value in

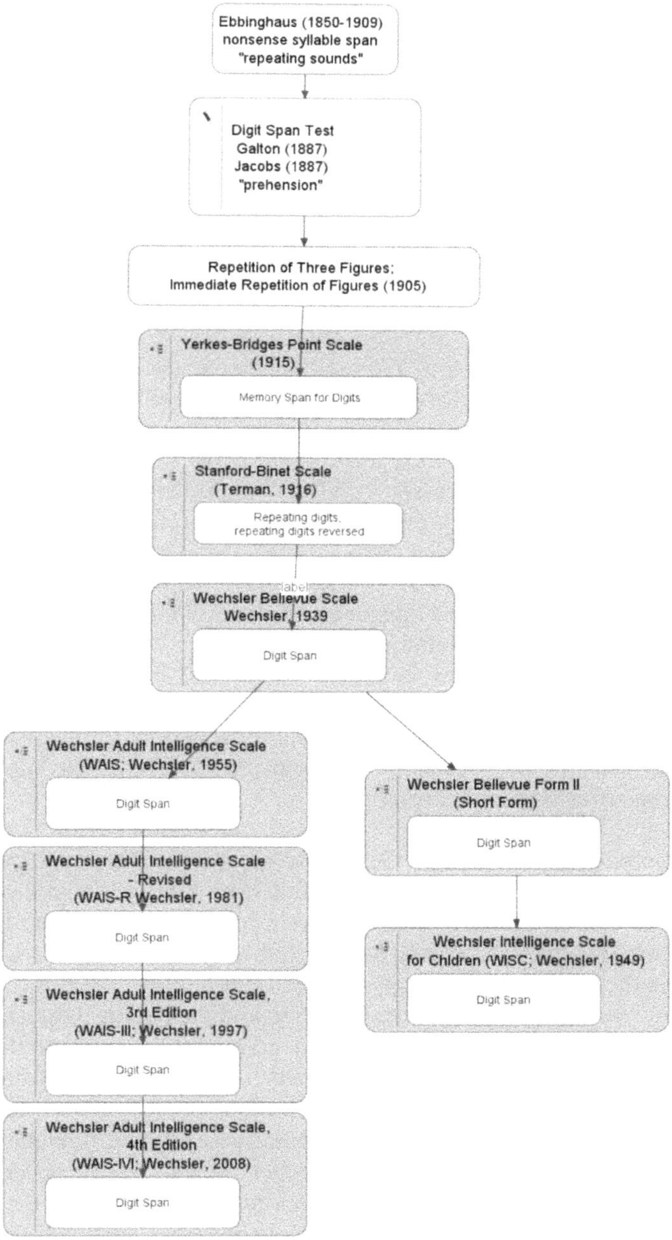

Figure 4. History of the "Digit Span" task (adapted from Boake, 2002), starting with the work of Ebbinghaus in the nineteenth century, through to the most recent release of the WAIS-IV in 2008. *Note:* Citation details are provided by Boake (2002).

determining how best to synthesise data for specific purposes. Particularly important is the fact that this taxonomic development is ideally suited for collaborative, incremental development. Such work might be facilitated by the forthcoming release of the American Psychological Association's PSYCTest database, which may offer a substantial target collection of psychological test names and associated annotation.

LESSONS FROM GENE ONTOLOGIES

The Gene Ontology (GO; www.geneontology.org) project stands as a major success story in ontology development, and provides a useful example for development of schemata to represent higher levels of phenotypic expression. GO started as an effort to foster better compatibility among databases containing information about three model organisms (worms, flies, and mice). It has grown to include more than 25,000 terms in three ontologies spanning: biological processes, cellular components and molecular functions. GO has emerged as a research standard completely through interest of participating scientists (rather than through any efforts to impose or enforce this standardisation). Bada et al. (2004) have emphasised a number of the key elements that fostered the successful development, deployment, and adoption of GO by its community.

Elements of this success included: community engagement, clear goals, limited scope, simple intuitive structure, continuous evolution, active curation, and early adoption. In the following list, we enumerate how each of these features of successful ontology development may apply to the development of cognitive ontologies.

- *Community engagement*: Gene ontologies benefited from an eager and active group of investigators interested in developing methods to cope with the burgeoning data that were emerging from high-throughput genotyping efforts. It was not created by external engineers, but rather emerged within the biology research community from the individuals who were facing challenges finding and managing large amounts of information. We have at hand a comparably complex if not more daunting challenge, and will need to gather input from multiple disciplines, including but not limited to: cognitive neuroscience, neuropsychology, behavioural genetics, and neuroimaging research communities. Different levels of expertise will also be needed. Certain features of cognitive ontology development will likely benefit from the input of highly experienced experts in the field. Particularly the initial postulates, assertions, and arguments about these may appropriately be generated by "opinion leaders" who possess extensive prior knowledge.

But the ultimate success in fleshing out relevant literature, detailing specific effects, and links to data will demand wide-ranging effort from individuals at all levels of training. Doctoral and postdoctoral trainees may be ideally suited to participate as critical stake holders given their engagement in digesting novel corpora of literature, and their efforts to develop coherent frameworks for digesting the evidence.

- *Clear goals and limited scope*: The GO project benefited from having goals limited to the specification of gene products so that there would be consistent annotation of biological databases. There is a risk that development of a cognitive ontology could founder on efforts to make it too comprehensive, and thus it will be important to limit the scope of concepts that will be target of cognitive ontology development. As noted earlier, specification of any ontology requires minimally a lexicon (the collection of terms that describe the entities contained in the ontology) and specification of relation types (the actual semantic descriptions of how entities relate to each other). The scope of the cognitive concept list need not be exhaustive for proof of concept work, but should be reasonably comprehensive in order to support early applications such as unbiased literature retrieval. The Consortium for Neuropsychiatric Phenomics and Cognitive Atlas projects at UCLA have generated lists of some 5000 concepts representing "cognition", and through automated and manual editing narrowed this lexicon to some 2000 terms that we believe offer a reasonable starting point in the representation of cognitive concepts and cognitive tasks. Specifying relations is complex, but there is likely a need to identify: (a) relations of cognitive concepts to specific task indicators (variables) used to measure these concepts (internal validity indicators); (b) relation of task indicators to the tasks themselves (it is possible to identify both general and task-specific context effects on these indicators; for example "response time" may have broad relations, across multiple tasks to some other external variables like aging, but also may mean markedly different things in one task relative to another); (c) relation of task indicators to other variables (external validity indicators). There may be many types of external validity indicators, but we believe it is useful to consider as starting points: (i) effect sizes for group difference (e.g., patients vs. controls) and (ii) correlations (e.g., test variable correlates with age or cortical grey matter volume; notably other measures of association with variables that are categorical may also be transformed conveniently into correlation coefficients). The CNP work on cognitive ontologies is targeting selectively cognitive concepts and the tasks used to measure these. Elaboration of these concepts and relations to other levels is currently kept minimalistic, but open for individual user annotations to input more elaborate qualifications, and these may suggest future directions for extension.

- *Simple intuitive structure*: GO benefited from the use of some simple, intuitively appealing class hierarchies and annotation links to decrease the barriers to use. The CNP and CA work on cognitive ontologies is also based on simple class hierarchies of cognitive concepts, and association graphs that help to illustrate how task indicators relate to the cognitive concepts, and how the cognitive concepts relate to each other. This fosters relatively straightforward navigation, and the capacity to browse readily at varying depths.
- *Early use, active curation, and continuous evolution*: Key elements in the success of GO included development from initial goals with stepwise, meaningful additions, active engagement of its initial authors, and a growing user base. It is too soon to tell what early use and adoption patterns will be for ontologies representing cognitive concepts and their interrelations. Usage patterns will clearly depend on many of the factors listed previously, particularly the perceived value to users for their work, and removing barriers to entry and use. Active curation will depend on adequate support, which will in turn likely depend on usage and perceived utility to the user community. Continuous evolution will depend on both use, diversity of inputs, and flexibility of the initial design, so that it readily accommodates change.

COGNITIVE ONTOLOGY DEVELOPMENT: FRAMEWORKS FOR LITERATURE MINING AND COLLABORATIVE KNOWLEDGE BUILDING

We have initiated several projects to help develop cognitive ontologies, which we hope ultimately will both foster increased understanding and sharing of conceptual frameworks important to cognitive neuroscience, and further enable connections of information about cognitive phenotypes to other levels of biological knowledge. Some of this development took place under the aegis of the Center for Cognitive Phenomics, which was supported by a planning grant from the NIH Roadmap Initiative programme in interdisciplinary research (P20RR020750). We are continuing this work within the Consortium for Neuropsychiatric Phenomics, particularly in the project "Hypothesis Web Development for Neuropsychiatric Phenomics" (RL1LM009833; D. Stott Parker, PI), and also in a new project "The Cognitive Atlas" (R01MH082795, Russ Poldrack, PI). Links to these projects and associated resources are maintained at www.phenomics.ucla.edu.

Among these developments are multiple freely available web applications. Several of these web-based applications support examination and visualisation of cognitive concepts in the scientific literature as this is represented in PubMed (PubGraph, PubAtlas, and PubBrain). The scientific literature is essentially the phenomics database we possess today. From a conceptual standpoint, this can be considered a very large database that is minimally structured, and thus requires new literature mining technology to extract useful information. The phenomics information it contains is complex—far more complex than genomics data. Genomics data are linked explicitly with sequences, and hence can use sequences as an indexing framework for retrieving information. Phenomics is more challenging because phenotypes involve diverse measurements that lack a natural organisation or indexing framework. Phenotype measurements in the literature are communicated using phrases, obliging us to retrieve information using simple language queries and other tools for literature mining. PubGraph, PubBrain, and PubAtlas are all attempts to provide "phenotype maps" based on the literature in PubMed/MEDLINE.

PubGraph allows a general specification of phenotypes, and maps phenotypes corresponding to individual literature queries into a topology (graph) reflecting strength of association. PubGraph permits users to either select from a number of preestablished lexica, or to input their own lexica, as the queries to PubMed. Nodes in these graphs reflect results of individual query expressions, and edges in these graphs represent measures of association between the nodes (e.g., the Jaccard coefficient, which is the intersection divided by the union of the queries).

PubAtlas emphasises scale, using a grid to map associations of larger sets of phenotypes, and it also supports development of lexica to formalise hierarchies of phenotypes in the same way that the BLAST sequence alignment tool uses databases to formalise related sets of genetic sequences. These lexica are themselves elementary ontologies. PubAtlas offers similar capabilities to PubGraph, but with additional features, including thresholding by association strength, along with "velocity plots" showing the cooccurrences history by year. For example, Figure 5 illustrates cooccurrences of the term "digit span" with a lexicon of 900 cognitive concept terms, thresholded to reveal only the strongest associations (natural log of Jaccard coefficient $> 10^{-4}$). The association table shows that the strongest associations (other than associations with "digit span" itself, are with "attention span" and "memory span"). Also shown in Figure 5 is the history plot for association of "digit span" with the concept "working memory", which reveals a steady increase in the late 1980s and an acceleration of use in the twenty-first century. By clicking on any of the table entries, PubAtlas invokes PubMed and generates the query of interest so that the user can inspect the relevant literature directly.

PubBrain maps the literature into phenotype space—by displaying PubMed hits in the 3-D phenotype geometry of brain anatomical regions.

Figure 5. PubAtlas output for the intersection of "digit span" with a lexicon of 900 cognitive concepts, thresholded to show only associations with ln Jaccard coefficient > -4.0. The "heat map" reveals strong associations with "memory span", "attention span", and "working memory". The insert shows the history in use of the term "working memory" together with "digit span".

PubBrain performs this using a "blast" query of any arbitrary PubMed search with respect to an ontology representing neuroanatomic terms (adapted from the Foundational Model of Anatomy and its NeuroNames

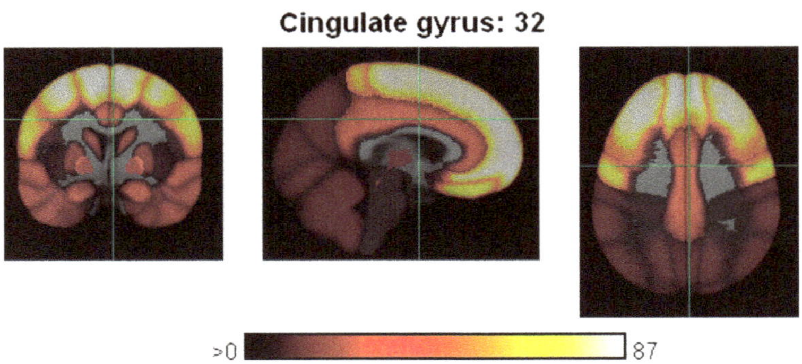

Figure 6. PubBrain output showing the projection of the query "digit span" on a three-dimensional probabilistic atlas of the brain; this view was centred on the cingulate gyrus (32 hits). Other frontal cortical regions had up to 87 cooccurrences.

lexicon). After identifying cooccurrences of a specific term with the anatomic terms, PubBrain then projects the cooccurrence statistics onto a three-dimensional probabilistic atlas of the human brain, which was developed in collaboration with the Laboratory of NeuroImaging (LONI) at UCLA (Shattuck et al., 2008). Figure 6 provides the PubBrain output for the query "digit span", illustrating the retrieval results for an arbitrary location (crosshairs positioned in the cingulate gyrus), and generating links to 32 publications that were used to generate this map.

These systems (PubGraph, PubBrain, and PubAtlas) illustrate how phenomics requires new tools centred on the phenotype concept. These tools now rest on PubMed, but they will be able to evolve as phenomics becomes better formalised, and curated phenomic databases become important. Thus, they not only show the importance of visual interfaces for understanding complex phenotype data, but also give us a view of "phenotype mapping" tools of the future.

Several other projects are underway to foster collaborative development of ontologies and collect annotation about cognitive concepts. The explosion of social collaborative networking and collaborative knowledge development during the "Web 2.0" era resulted in some dramatic achievements. The Wikipedia is one of the most widely appreciated of these deliverables, now containing more than 2.5 million articles. Wikipedia already is almost 40 times larger than the Encyclopaedia Britannica, and at least some investigations have suggested the quality of scientific articles is not markedly different between these two sources of information (Giles, 2005). Concern about the validity of Wikipedia's content is not preventing its widespread use and continued growth. Only a decade ago "the wisdom of crowds" might have seemed an oxymoron, but today there is concrete evidence that Web 2.0 has fulfilled at least part of its promise. The development of the web itself spawns futuristic thinking, and Web 3.0 already has been a topic of discussion since at least 2006 (this coinage is attributed to John Markoff of the *New York Times*, according to Wikipedia). According to some, the way forward will be paved with key features of the *semantic web*, which would add semantic structure and conventions to web contents to facilitate information storage, retrieval, and sharing. Ontology development is among the key elements of the Web 3.0 vision. In both the Consortium for Neuropsychiatric Phenomics and Cognitive Atlas projects, collaborative knowledge building is a central goal.

The Phenowiki (www.phenowiki.org) was developed to serve as a collaborative database for phenotype annotation. It contains wiki-like descriptions for a series of cognitive tasks and concepts, and these task and concept entries are linked to an underlying database that contains quantitative annotation about specific study results. For example, it already includes annotations about eight publications that used the Digit Span test, including quantitative results about heritability coefficients, effect sizes for group differences between various patient and comparison groups, and reliability coefficients for the specific measures that are derived for this test. Some additional applications of the Phenowiki were described by Sabb et al. (2008). The Phenowiki already can serve as a collaborative repository for storing quantitative effect size information about group differences, treatment effects, and other psychometric properties of individual tasks, and outputs of the Phenowiki can be used directly to generate meta-analytic summary statistics. Future developments of this tool are aimed to enable automated meta-analysis of selected topic areas and for specific task results.

The Hypothesis Web project aims further to develop collaborative tools for the specification, visualisation, and sharing of scientific hypotheses with quantitative annotation. Currently, most hypotheses are laid out in scholarly discussions as a series of verbal postulates without much quantification of each assertion. Instead, each assertion is assigned a dichotomous value (i.e., "true") based on scientific findings that are almost always only true to some degree of statistical certainty. We might take for example the "COMT hypothesis of schizophrenia" (this example chosen because at least one of us [RMB] is guilty of such unquantified theoretical speculation). The original hypothesis linking the Val^{158}Met polymorphism in the COMT gene with schizophrenia included some tantalising elements: the COMT gene affects the COMT enzyme; the COMT enzyme affects dopamine (DA) metabolism; antipsychotic drugs affect DA metabolism; antipsychotic drugs ameliorate symptoms of schizophrenia; DA metabolism affects frontal function and cognitive function; velocardiofacial syndrome involves a "knockout" of the COMT gene and is associated with schizophrenia; and more (for a review, see Williams, Owen, & O'Donovan, 2007). This hypothesis has now been the subject of more than 250 papers, which "virtually exclude" the likelihood of a simple link between the most studied Val^{158}Met polymorphism and schizophrenia (Williams et al., 2007). Imagine that from the beginning of this quest to find links of a genetic variant to a complex syndrome such as schizophrenia, we had conditioned each assertion with clearer explication of quantitative effect sizes (and appropriate confidence intervals) linking each level of this hypothesis to the next. By stating more explicitly that the COMT gene has only partial effect on the enzyme, the enzyme only partial effect on DA concentrations in the PFC, DA concentrations in PFC only partial effect on neural network function, and so on, our failure to find associations with the complex syndrome might seem less surprising, and help keep us focused on the critical dimensions of the hypothesis that may possess validity and be important for cognitive neuroscience. The Hypothesis Web project aims to bring to users convenient methods for specifying and manipulating assertions, with quantitative annotation, in a collaborative framework. Although still early in its development, the PubAtlas application already enables users to generate literature-based "models" that specify the nodes of a particular hypothesis as articulated via PubMed queries, and to assess the associations between nodes in terms of literature cooccurrence statistics. It is hoped that one day applications like the Hypothesis Web will help revolutionise the way we present hypotheses (for example, in the Background and Significance sections of grant applications, and in the Introduction sections of scientific publications). Rather than presenting arguments as a series of partially qualified statements, hypotheses of the future may include more formal specification and quantification of their evidentiary bases.

The Cognitive Atlas project specifically aims to provide the infrastructure for collaborative development of cognitive ontologies. The central organising framework for this development will be a web application in which cognitive concepts are represented. It will allow networks of researchers to collaborate on assertions about mental concepts, including how these are measured, how the concepts relate to each other, and how the concepts relate to other entities (such as diagnostic groups, drug effects, or neural system activations). A key feature of this infrastructure will be its capacity to represent uncertainty about assertions, and to represent explicitly disagreement and controversy. This will involve verbal arguments for and against key assertions, but will further enable representation of quantitative evidence supporting the validity (or lack thereof) regarding assertions. For example, a specific "node" in the cognitive atlas representing the validity of a concept ("the phonological buffer") might be supported by some literature (Baddeley & Hitch, 1974) but not supported by other literature (Jones, Macken, & Nicholls, 2004). Users will be able to "vote" as to which of these two contradictory opinions they find most persuasive, providing a quantitative index of community-perceived validity, and add comments supporting their opinions. Additional quantitative indices regarding cognitive constructs will be provided by associating empirical data with the edges in the conceptual network. For example, the assertion that "working memory is impaired in schizophrenia" might be supported by evidence showing that patients with the diagnosis of schizophrenia showed deficits relative to a healthy comparison group on the "N-back task", and that this relative deficit is associated with a quantitative effect size (such as Cohen's *d*), of a specific magnitude, and with quantified confidence intervals around that effect size. With the accrual of a sufficient database of empirical findings, it should become possible to test competing ontologies by comparing their implied covariance structure to the actual covariance structure of these data, using methods like the meta-analytic structural equation modelling approach described earlier. The knowledge base developed in the Cognitive Atlas will provide a formal conceptual infrastructure for the annotation of behavioural and neuroscientific data, allowing the kind of large-scale data mining that has become central in bioscience but has not yet penetrated cognitive neuroscience.

SUMMARY AND CONCLUSION

We now face unprecedented challenges and opportunities to pursue the new transdiscipline of phenomics and relate the human genome to complex neuropsychiatric syndromes. Cognitive phenotypes may offer a rational focus of inquiry and help bridge knowledge from neural systems to symptomatic levels. To enable these connections cognitive neuropsychiatry needs *cognitive ontologies*—structured representations of cognitive concept domains and their measurements. This paper summarises some of the specific challenges faced in developing cognitive ontologies, and offers strategies to help overcome barriers posed by the abstract nature and temporal instability of cognitive concepts. Examples of works-in-progress and opportunities for collaborative development are presented. It is hoped that these efforts will help advance cognitive neuroscience and ultimately our understanding and treatment of neuropsychiatric syndromes.

REFERENCES

American Psychiatric Association (1994). *Diagnostic and Statistical Manual of Mental Disorders (DSM-IV).* Washington, DC: Author.

Bada, M., Stevens, R., Goble, C., Gil, Y., Ashburner, M., Blake, J. A., et al. (2004). A short study on the success of the Gene Ontology. *Journal of Web Semantics, 1*, 235–240.

Baddeley, A. D., & Hitch, G. (1974). Working memory. In G. H. Bower (Ed.), *The psychology of learning and motivation: Advances in research and theory* (Vol. 8 pp. 47–89). New York: Academic Press.

Bilder, R. M. (2008). Phenomics: Building scaffolds for biological hypotheses in the post-genomic era. *Biological Psychiatry, 63*, 439–440.

Bilder, R. M., Volavka, J., Czobor, P., Malhotra, A. K., Kennedy, J. L., Ni, X., et al. (2002). Neurocognitive correlates of the COMT Val(158)Met polymorphism in chronic schizophrenia. *Biological Psychiatry, 52*, 701–707.

Bilder, R. M., Volavka, J., Lachman, H. M., & Grace, A. A. (2004). The catechol-O-methyltransferase polymorphism: Relations to the tonic-phasic dopamine hypothesis and neuropsychiatric phenotypes. *Neuropsychopharmacology, 29*, 1943–1961.

Boake, C. (2002). From the Binet-Simon to the Wechsler-Bellevue: Tracing the history of intelligence testing. *Journal of Clinical and Experimental Neuropsychology, 24*, 383–405.

Braver, T. S., & Barch, D. M. (2002). A theory of cognitive control, aging cognition, and neuromodulation. *Neuroscience and Biobehavioral Reviews, 26*, 809–817.

Centler, F., Kaleta, C., di Fenizio, P. S., & Dittrich, P. (2008). Computing chemical organizations in biological networks. *Bioinformatics, 24*, 1611–1618.

Cheung, M. W. (2008). A model for integrating fixed-, random-, and mixed-effects meta-analyses into structural equation modeling. *Psychological Methods, 13*, 182–202.

Chiang, P. K., Lam, M. A., & Luo, Y. (2008). The many faces of amyloid beta in Alzheimer's disease. *Current Molecular Medicine, 8*, 580–584.

Egan, M. F., Goldberg, T. E., Kolachana, B. S., Callicott, J. H., Mazzanti, C. M., Straub R. E., et al. (2001). Effect of COMT Val108/158 Met genotype on frontal lobe function and risk for schizophrenia. *Proceedings of the National Academy of Sciences of the USA*, *98*, 6917–6922.

Ferreira, M. A., O'Donovan, M. C., Meng, Y. A., Jones, I. R., Ruderfer, D. M., Jones, L., et al. (2008). Collaborative genome-wide association analysis supports a role for ANK3 and CACNA1C in bipolar disorder. *Nature Genetics, 40*, 1056–1058.

Flint, J., & Munafo, M. R. (2007). The endophenotype concept in psychiatric genetics. *Psychological Medicine*, *37*, 163–180.

Frayling, T. M. (2007). Genome-wide association studies provide new insights into type 2 diabetes aetiology. *Nature Reviews: Genetics*, *8*, 657–662.

Freimer, N., & Sabatti, C. (2004). The use of pedigree, sib-pair and association studies of common diseases for genetic mapping and epidemiology. *Nature Genetics*, *36*, 1045–1051.

Freimer, N. B., & Sabatti, C. (2005). Guidelines for association studies in Human Molecular Genetics. *Human Molecular Genetics*, *14*, 2481–2483.

Furlow, C. F., & Beretvas, S. N. (2005). Meta-analytic methods of pooling correlation matrices for structural equation modeling under different patterns of missing data. *Psychological Methods*, *10*, 227–254.

Fuster, J. M., & Alexander, G. E. (1971). Neuron activity related to short-term memory. *Science*, *173*, 652–654.

Giles, J. (2005). Internet encyclopaedias go head to head. *Nature*, *438*, 900–901.

Goldman-Rakic, P. S. (1987). Circuitry of primate prefrontal cortex and regulation of behavior by representational memory. In V.B. Mountcastle (Ed.), *Handbook of physiology: Vol. 5. The nervous system* (pp. 373–417). Bethesda, MD: American Physiological Society.

Grethe, J., Gupta, A., Ludascher, B., & Martone, M. E. (2003). BIRN ontologies session. Vol. 2008 [PowerPoint slide show].

Gruber, T. R. (2003). A translation approach to portable ontologies. *Knowledge Acquisition*, *5*, 199–220.

Hariri, A. R., Goldberg, T. E., Mattay, V. S., Kolachana, B. S., Callicott, J. H., Egan, M. F., & Weinberger, D. R. (2003). Brain-derived neurotrophic factor val66met polymorphism affects human memory-related hippocampal activity and predicts memory performance. *Journal of Neuroscience*, *23*, 6690–6694.

Haslam, N. (2003). Categorical versus dimensional models of mental disorder: The taxometric evidence. *Australian and New Zealand Journal of Psychiatry*, *37*, 696–704.

Haslam, N., & Kim, H. C. (2002). Categories and continua: A review of taxometric research. *Genetic Social and General Psychology Monographs*, *128*, 271–320.

Jacobsen, C. F. (1936). Studies of cerebral function in primates: I. The function of the frontal association areas in monkeys. *Journal of Comparative Psychology*, *13*, 3–60.

Jones, D. M., Macken, W. J., & &Nicholls, A. P. (2004). The phonological store of working memory: Is it phonological and is it a store? *Journal of Experimental Psychology: Learning, Memory, and Cognition*, *30*, 656–674.

Kauffman, S. (1995). *At home in the universe: The search for laws of self-organization and complexity*. New York: Oxford University Press.

Kendler, K. S. (2005). Psychiatric genetics: A methodologic critique. *American Journal of Psychiatry*, *162*, 3–11.

Kraemer, H. C. (2007). DSM categories and dimensions in clinical and research contexts. *International Journal of Methods in Psychiatric Research*, *16*, S8–S15.

Miller, E. K., & Cohen, J. D. (2001). An integrative theory of prefrontal cortex function. *Annual Review of Neuroscience*, *24*, 167–202.

Nolan, K. A., Bilder, R. M., Lachman, H. M., & Volavka, J. (2004). COMT Val158Met polymorphism in schizophrenia: Differential effects of Val and Met alleles on cognitive stability and flexibility. *American Journal of Psychiatry*, *161*, 359–361.

O'Donovan, M. C., Craddock, N., Norton, N., Williams, H., Peirce, T., Moskvina, V., et al. (2008). Identification of loci associated with schizophrenia by genome-wide association and follow-up. *Nature Genetics, 40*, 1053–1055.

Poldrack, R. A. (2006). Can cognitive processes be inferred from neuroimaging data? *Trends in Cognitive Science, 10*, 59–63.

Posner, M. I., & Snyder, C. R. R. (1975). Attention and cognitive control. In R. L. Solso (Ed.), *Information processing and cognition* (pp. 55–85). Hillsdale, NJ: Lawrence Erlbaum Associates, Inc.

Pribram, K. H., & McGuinness, D. (1975). Arousal, activation, and effort in the control of attention. *Psychological Review, 82*, 116–149.

Riley, R. D., Abrams, K. R., Sutton, A. J., Lambert, P. C., & Thompson, J. R. (2007). Bivariate random-effects meta-analysis and the estimation of between-study correlation. *BMC Medical Research Methodology, 7*, 3.

Sabb, F. W., Bearden, C. E., Glahn, D. C., Parker, D. S., Freimer, N., & Bilder, R. M. (2008). A collaborative knowledge base for cognitive phenomics. *Molecular Psychiatry, 13*, 350–360.

Saura, C. A., Chen, G., Malkani, S., Choi, S. Y., Takahashi, R. H., Zhang, D., et al. (2005). Conditional inactivation of presenilin 1 prevents amyloid accumulation and temporarily rescues contextual and spatial working memory impairments in amyloid precursor protein transgenic mice. *Journal of Neuroscience, 25*, 6755–6764.

Shattuck, D. W., Mirza, M., Adisetiyo, V., Hojatkashani, C., Salamon, G., Narr, K. L., et al. (2008). Construction of a 3D probabilistic atlas of human cortical structures. *Neuroimage, 39*, 1064–1080.

Sporns, O., Tononi, G., & Kotter, R. (2005). The human connectome: A structural description of the human brain. *PLoS Computational Biology, 1*, e42.

Sridhar, S., Dhamdhere, K., Blelloch, G., Halperin, E., Ravi, R., & Schwartz, R. (2007). Algorithms for efficient near-perfect phylogenetic tree reconstruction in theory and practice. *IEEE/ACM Transactions in Computational Biology and Bioinformation, 4*, 561–571.

Tononi, G., Sporns, O., & Edelman, G. M. (1994). A measure for brain complexity: relating functional segregation and integration in the nervous system. *Proceedings of the National Academy of Sciences of the USA, 91*, 5033–5037.

Tononi, G., Sporns, O., & Edelman, G. M. (1999). Measures of degeneracy and redundancy in biological networks. *Proceedings of the National Academy of Sciences of the USA, 96*, 3257–3262.

Williams, H. J., Owen, M. J., & O'Donovan, M. C. (2007). Is COMT a susceptibility gene for schizophrenia? *Schizophrenia Bulletin, 33*, 635–641.

Zheng, Y., Yeh, C. W., Yang, C. D., Jang, S. S., & Chu, I. M. (2007). On the local optimal solutions of metabolic regulatory networks using information guided genetic algorithm approach and clustering analysis. *Journal of Biotechnology, 131*, 159–167.

COGNITIVE NEUROPSYCHIATRY
2009, 14 (4/5), 451–472

Psychology Press
Taylor & Francis Group

Phenotyping psychosis: Room for neurocomputational and content-dependent cognitive endophenotypes?

Eric Y. H. Chen, Gloria H. Y. Wong, Christy L. M. Hui, Jennifer Y. M. Tang, Cindy P. Y. Chiu, May M. L. Lam, and P. C. Sham

Department of Psychiatry, University of Hong Kong, Hong Kong

Introduction. The endophenotype research strategy aims at reducing complex clinical phenomena to reveal a more tractable mapping to underlying genes. Cognitive dysfunctions have been widely pursued as target endophenotype in schizophrenia. We critically discuss the promise and limitations of this approach.
Methods. Relevant theoretical and empirical issues on genes and behaviour, neurocognitive structure and psychopathology were selectively reviewed and discussed.
Results. Some important inherent limitations of the current cognitive endophenotype approach were identified. These include reliance on (1) classic neuropsychology; (2) deficit measures; and (3) a general information processing approach with the use of content-independent, neutral stimuli. As a result, many current cognitive endophenotypes are likely to overlap and converge with general cognitive impairments, which may be shared with other disorders.
Conclusions. We propose three novel directions for further psychosis endophenotype research: (1) in addition to such content-independent computational processes, which operate in a similar way regardless of the stimuli, it is important to consider the potential roles of "content-dependent endophenotypes", which operate on different stimuli in consistently different manners. Advances in cognitive studies suggest there may be evolutionarily important aspects of cognition which are content-dependent. We propose that both content-independent and content-dependent processes should be addressed in psychosis research. (2) In line with the emphasis on content, close attention should be paid to the study of "psychopathological endophenotypes" in addition to cognitive endophenotypes. (3) "Neurocomputational endophenotypes" may be defined by parsing cognitive processes into "subsystems" with specific computational processing algorithms and considering key computational parameters suggested from these models. These potential "neurocomputational endophenotypes" (such as neuronal noise, synaptic

Correspondence should be addressed to Eric Y.H. Chen, Department of Psychiatry, University of Hong Kong, Queen Mary Hospital, Pokfulam Road, Hong Kong. E-mail: eyhchen@hku.hk

http://www.psypress.com/cogneuropsychiatry DOI: 10.1080/13546800902965695

learning algorithms) are potentially intermediate variables located between the levels of cognition and neurobiology.

Keywords: Psychosis; Endophenotype; Genotype; Cognitive structure; Content-dependent process.

INTRODUCTION: WHY ENDOPHENOTYPING IS IMPORTANT

Definition of the target behavioural phenotype is a crucial starting point in any attempt to map genes to behaviours. In the case where the target behavioural phenotype is the complex disorder schizophrenia, data so far suggest a polygenic mode of inheritance with each of the putative genes exerting only a small effect (Harrison & Owen, 2003). There are likely many non-linear gene–gene and gene–environment interactions (Gottesman & Shields, 1967; Meyer-Lindenberg & Weinberger, 2006), which may be further complicated by possible pleiotropic effects (Page, George, Go, Page, & Allison, 2003). This challenging scenario has led investigators to reconsider whether complex clinical categories (such as schizophrenia or psychotic disorders) are optimal phenotypes. As a result, researchers have attempted to reduce complex phenotypes to simpler endophenotypes in the hope that the latter represents putative "atomic" components which map more directly onto genes (Gottesman & Gould, 2003; Gould & Gottesman, 2006).

We critically evaluate this approach, particularly in relation to cognitive and psychopathological (as in descriptive psychopathology) phenotypes. After surveying the explanatory gap between genes and behaviour first from a bottom–up and then a top–down perspective, we propose that existing cognitive endophenotype research may have suffered from a relative lack of robust *a priori* considerations of search strategies and consequently may bias towards measuring general cognitive dysfunctions, resulting in likely findings of genetic mapping to nonspecific general cognitive capacity. While useful to some extent, this approach may not fully address the more specific pathways for psychosis. We then explore some current views on the structure of cognition in terms of cognitive subsystems and basic computational processes within a cognitive subsystem. This leads to the proposal of locating endophenotypes at the neurocomputational level, intermediate between cognition and neurobiology. At the same time we explore possible reasons why there has been a tendency to overlook content-dependent endophenotypes. This tendency was also apparent in empirical cognitive studies and psychopathology. Following works that suggest human cognitive processing may consist of content-specific modules, we propose that it is important to take into account "content-dependent" processes (different stimuli are processed in consistently different manners by the same processing system) as potential endophenotypes in addition to "content-independent" processes (different stimuli being processed by the same operation). It is argued that while content-independent processes may partly underlie psychotic disorders, they are less likely to be specific. It is therefore proposed that it will be the content-dependent processes, which have been relatively neglected in traditional cognitive research, that are most promising in yielding endophenotypes relevant to a specific understanding of psychosis.

THE GAP BETWEEN GENOTYPE AND PHENOTYPE

Recent progresses in genotyping and sequencing technologies have resulted in a situation whereby gene-level information has outstripped our knowledge of how genes work at physiological and pathological levels. This discrepancy is particularly noted for brain mechanisms in relation to complex behavioural disorders.

Complexity of the "black box": bottom–up view

Current knowledge on processes that explain from the bottom–up how genes are expressed in behaviour is minimal. Still it is important to try to appreciate the complexity involved. Amongst possible direct expressions of genes, some may specify structural proteins in the nervous system (e.g., the axon and cytoskeleton dendritic trees) and some specify chemical transmission processes in synapses (e.g., receptors, neurotransmitters, second messengers, and metabolism of neurotransmitters). Apart from the structural genes, there may be genes that specify the physical location of particular cells, as well as those that mark specific temporal stages along development. Some genes may mediate specific environmental signals (e.g., the mating season, seasons in general, and food scarcity). Some genes may even specify fixed connectivity between neurons (e.g., instinctive perceptual patterns and instinctive sets of responses). There are also regulatory genes that directly modulate the expression of other genes; those that can be activated depending on the internal environment (i.e., responding to other components of the neural system) and those that respond to aspects of the external environment. Together, genes of the brain (estimated to comprise approximately a third of the human genome) drive the formation of different cell types, the axonal growth and tracking, the formation of neural connections, as well as the connection pattern between neurons (i.e., networks) and the timing of neural developmental events. Eventually after many levels of reduction, they constitute the material basis for the behavioural modules, with input–output mappings modulated by a large number of factors.

Challenges of the "black box": top–down view

The use of endophenotypes are intended as an initial top–down deconstruction (i.e., reduction) of complex behavioural phenomena into component processes (i.e., cognitive or neural), which presumably lie closer to the biological mechanisms that genetic information impacts on.

Currently there are at least three levels of reduction: the symptom/psychopathology level (i.e., units of subjective experience/behavioural categories); the cognition level (i.e., discrete cognitive processes or modules); and the neural level (i.e., brain physiological systems). It is implicitly assumed that these levels are hierarchically arranged so that phenomena at a higher level are reducible to combinations of "atomic" components at a lower level. For example, a symptom may be reducible to one or more basic cognitive processes, each of which in turn is reducible to one (or more) brain process(es). It is assumed that lower level "atomic" components are more directly determined by a smaller number of genes. There is however no clear guidance as to what might constitute the "natural kinds" or whether we are "carving nature at its joints" at lower levels. Whilst cognitive science has attempted to define the structure of cognition in the form of "cognitive modules and subsystems" (e.g., Kosslyn, 1994; Pylyshyn, 1984), it has been challenging to consider whether such conceptions are likely to be relevant in the search for psychosis endophenotypes (see later).

LIMITATIONS IN CURRENT COGNITIVE ENDOPHENOTYPE APPROACHES

Power to reduce complexity from clinical phenotypes

As endophenotypes are intended to be alternative behavioural dimensions that map better to genes than clinical categories (Gottesman & Gould, 2003), their success should be measured against this objective. It is contentious as to whether so far the application of cognitive endophenotypes to understanding psychosis specifically has resulted in a definitive reduction in complexity (Flint & Munafo, 2007; Keri & Janka, 2004).

Figure 1 illustrates two potential scenarios with endophenotypes. In Figure 1a, depicting the ideal situation, a clinical phenotype (diagnostic category) is completely reduced to two distinct endophenotypes. Each of these neatly maps onto a smaller number of genes. In this instance, the use of endophenotype has reduced the complexity of the genotype–phenotype mapping and made the subsequent analysis simpler. In another (more likely) situation, depicted by Figure 1b, the phenotypes are (1) not completely explained by the endophenotypes, and (2) the endophenotypes are not specific for the phenotypes. In this situation it is possible that the endophenotypes

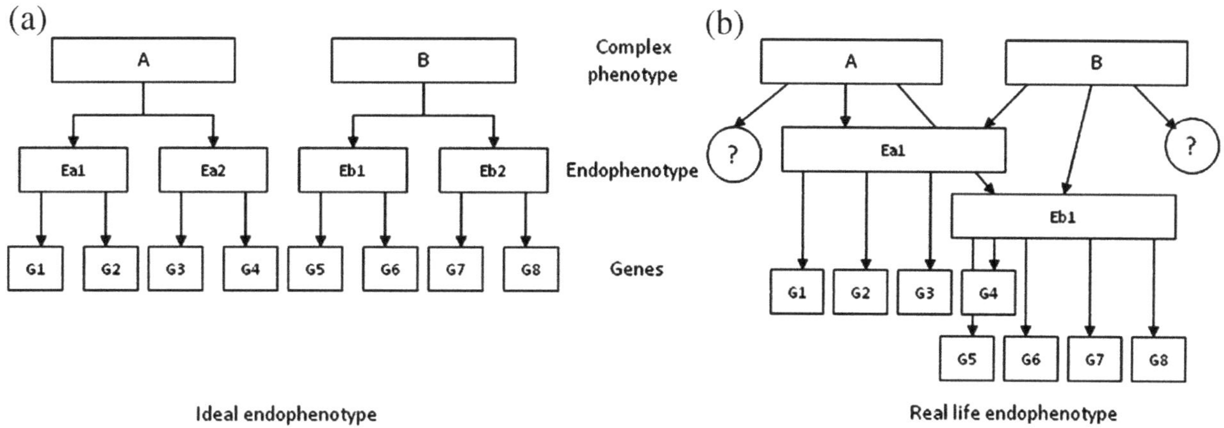

Figure 1. Schematic representation of ideal and real-life endophenotypes mappings. A and B Represent phenotypes (diagnostic categories). Ea1, Ea2, Eb1, Eb2 are putative endophenotypes (cognitive or biological measures). G1-8 are related genes. Figure 1a is the ideal situation when the endophenotype-to-gene mapping is less complex than the phenotype-to-gene mapping. Figure 1b is the real-life situation where the endophenotype-to-gene mapping may not be less complex as compared with the phenotype-to-gene mapping.

themselves may map onto an equally large set of genes, and not necessarily significantly reduce the complexity of the phenotype-to-gene mapping. Arguably the current situation for schizophrenia may be closer to the latter than the former instance.

Endophenotypes are "endo" because they may not be easily observable (John & Lewis, 1966). In the search for cognitive endophenotypes, dimensions that are relatively more accessible have been preferentially studied. Existing screening for potential cognitive endophenotypes has relied heavily on a convenient set of currently available measurements (Goldberg & Weinberger, 2004) – pre-existing assessment tools for memory, attention, executive function and general intelligence in psychosis dominated recent research, without necessarily providing richer insights on how the failure of these systems mediates psychotic symptoms. A fundamental limitation of current research is thus the lack of a coherent strategy to guide researchers in looking beyond what is conveniently available.

General (content-independent) cognition and psychosis

Although psychotic illnesses express themselves in very specific terms (i.e., in the highly specific range of symptoms, such as persecutory delusions, derogatory third-person auditory hallucinations), there is also evidence that a general reduction in cognitive capacity may contribute towards psychosis. Psychotic symptoms have been reported more frequently in populations suffering from a general reduction in cognitive processing capacities (e.g., learning disability, dementia and ageing) (Rosen & Zubenko, 1991; Turner, 1989; van Os, Howard, Takei, & Murray, 1995). Psychotic symptoms can also occur in acute illnesses that compromise brain function integrity (e.g., traumatic brain injury) (Koponen et al., 2002). Research in cognitive dysfunctions in schizophrenia has suggested a contribution of general cognitive impairment (Kraus & Keefe, 2007). In fact, there has been intense debate over whether there are specific impairments on top of the generalized impairments in cognitive functions (e.g., working memory, verbal memory, executive function) (Goldberg, Weinberger, Berman, Pliskin, & Podd, 1987; Goldman-Rakic, 1994; Saykin et al., 1994). Regardless of the specific cognitive impairments, there is broad agreement that a generalized component of a wide range of deficits in cognitive function is involved (Dickinson, Ragland, Calkins, Gold, & Gur, 2006; Kraus & Keefe, 2007). It has been suggested that general cognitive abilities are determined by a set of genes (the "generalist" genes) (Butcher, Kennedy, & Plomin, 2006; Kovas & Plomin, 2006). Because of the relationship between general intelligence and psychosis, we anticipate that some of these genes may eventually be found to be associated with psychotic disorders.

How general cognitive impairment could contribute to psychosis is still unknown. A number of theories have been proposed. Some examples are: (1) general cognitive capacity is related to reasoning and logical thinking, and psychosis is a consequence of failure in applying logic and sound reasoning in interpreting social events (MacCabe et al., 2008); and (2) reduced cognitive capacity increases the likelihood of spurious memories generated by the associative memory system, resulting in the clustering of bizarre features on a single representation (Chen, 1994, 1995; Hoffman, 1987).

Even though it appears that nonspecific cognitive processes play a role in psychotic disorders, it is unlikely that they are sufficient explanations. It is important also to note that in psychosis, generalized cognitive impairment is neither a necessary nor a sufficient condition: a significant proportion (over 20%) of psychosis patients are not cognitively compromised (on classical neuropsychological measures), while a significant proportion of cognitively compromised subjects do not experience psychotic symptoms (Keri & Janka, 2004). There is considerable overlap between the cognitive capacity of participants with psychosis and normal participants. This highlights the need to further identify cognitive and neurobiological processes specifically related to the mechanisms of psychosis.

Evidence from twin studies also suggest in addition to a genetic link for more general neuropsychological measures and schizophrenia, there are also additional genetic factors unrelated to these measures that influence the risk of schizophrenia via independent mechanisms (Toulopoulou et al., 2007). It is in the search for these specific cognitive endophenotypes that a novel approach to characterize cognition in psychosis, distinct from the deficit and the general information processing frameworks, holds promise in identifying more specific cognitive pathways to psychosis.

Why conventional cognitive endophenotypes may be limited?

Most of the cognitive endophenotypes considered so far have originated from conventional neuropsychological measures or from a cognitive psychology framework. There is a possibility that they may converge to tap general cognitive impairment rather than become candidates for more specific psychosis endophenotypes, for the following reasons:

1. *Reliance on classic neuropsychology:* conventional neuropsychological measurements based on gross brain lesions (e.g., head injury, tumour or cerebrovascular events) appear to have played a major role in the search for cognitive endophenotypes. Although readily measurable, there is no *a priori* reason why such deficits should map more simply onto genes.

As most neuropsychological functions were defined by lesion effects on specific areas of the brain (such as the frontal lobe), this approach does less well in describing more diffused effects affecting networks of multiple brain areas. Some more sophisticated studies using modern cognitive science methods may overcome this limitation.

2. *Reliance on deficit measures:* a second limitation with the conventional cognitive approach is the emphasis on deficit measurements. The standard paradigm is to define a cognitive function and to measure the decline of that function in abnormal states. This approach is structurally less suitable for addressing problems that arise from aberrant functions rather than losses of function. For instance, conventional memory measurements focus on the loss of memory (e.g., failure to recognize and recall items in working memory, verbal and spatial long-term memory), and have relatively neglected the study of spurious memories (e.g., the production of abnormal or distorted memories). This bias would result in research findings that focus on general losses of functions, but would be relatively inadequate in probing the mechanisms for anomalous experiences in psychotic disorders.
3. *Reliance on a general information processing framework:* another weakness arises from the alignment of cognitive psychology with a general information processing framework (Tooby & Cosmides, 1992). The general information processing approach views the human cognitive system as a general purpose computational system, which processes different inputs to the system according to the same, neutral, algorithm independent to the content of the input (content-independent). As pointed out below, this approach may overlook important aspects of human cognition and psychopathology specifically relevant to psychosis. This content-independent framework could at best address only issues pertaining to general processing capacity and the disorders that arise from a nonspecific failure of processing, which are likely to be shared amongst many different disorders. Some investigators have already commented on the need for more novel perspectives in searching for cognitive endophenotypes (Goldberg & Weinberger, 2004). We would like to propose several possibilities in the rest of this paper.

PROBING THE STRUCTURE OF COGNITION

Cognitive subsystems and modularity

Applying the endophenotype approach to cognition raises a need for a rigorous review of the structure of cognition (i.e., how "individual" cognitive processes are validly defined, and how they are likely to be organized in relation to one another). Cognitive scientists have made several attempts at more rigorous conceptualizations of the structure of cognition. A commonly adopted framework conceives the cognitive system as consisting of a number of "subsystems". Each of the subsystem can receive an input, perform an operation on the input to generate an output, and then pass the output onto another subsystem. Operations within a subsystem are proposed to be relatively independent of processes occurring in other subsystems (e.g., Kosslyn, 1994). The concept of subsystems is akin to the notion of "cognitive modules" proposed by Fodor (1983). As such, these "cognitive modules" or "subsystems" naturally become candidates for the endophenotype approach, and an important task involves the identification and characterization of the key "subsystems" relevant to normal psychology as well as to psychosis. It is recognized however that the structure of cognition is unlikely to consist simply of a number of independent modules. Upon further consideration, the relationship between subsystems can be elaborated by a distinction between "strong modularity" and "weak modularity". In a metaphor used by Simon (1981) and adapted by Kossyln (1994), the cognitive system is compared to a house that is subdivided into rooms by walls (subsystems with strong modularity), and each room is subdivided into cubicles by thin walls (weak modularity). To regulate the room temperature there is a thermostat in each cubicle which can be set independently. This will result in each cubicle having different temperatures; however the temperatures in different cubicles within a room are more similar than that between rooms. Although such elaborations are important for retaining explanatory power over a wide range of phenomena, at the same time, they remind us that it may be simplistic to expect cognitive subsystems be "clean" endophenotypes.

Computation within cognitive subsystems

It is important to recognize that the conceptualization of cognition in terms of subsystems is only a first step. After defining the cognitive systems with their input and output, the crucial question is to understand how information is computed from input to output, and how information from one subsystem interacts with other subsystems. The interaction between subsystems can also take a number of forms: one subsystem may send output directly as input to another system, a subsystem's output might "prime" the processing in another subsystem so that it modulates the probability of a certain output in the target system; a subsystem may also affect the general operation of another subsystem (slowing or blocking it) (Kossyln, 1994). For the current discussion, we shall focus on computations within a cognitive subsystem.

Different accounts of cognitive computation exist, cognitive science makes a distinction between "symbolic" processes and "subsymbolic" processes (Smolensky, 1995). In symbolic processes, the input information is taken as an atomic unit, and the operation is defined at this level without further reduction (e.g., "STORE" stimulus "A", or "RETRIEVE" stimulus "A", thus maintaining a "black box" at this level). In "subsymbolic" processes the input information is further reduced to "dimensions" or "micro-features" (which may or may not be meaningful), and they are represented by multiple smaller units. Transformation from input to output is defined by parallel computation between these units using principles of associative learning. We shall consider one account of subsymbolic mental representation and operations (i.e., parallel distributed processing (PDP) or neurocomputational models). These models have been influential firstly because they adopt neurobiologically inspired basic structures: employing a large number of richly interconnected processing units (Rumelhart & McClelland, 1986). Secondly they offer a reduced level (sub-atomic level) of explanation compared with most conventional cognitive psychology models (typically articulated at the "symbolic" level with processes defined at this level without further reduction). Because of their unique roles, it is important to clarify the nature of neurocomputational models and the extent they impact on our thinking of endophenotypes. While most neurocomputational models are not aiming for detailed neuronal realism, and are certainly not complete explanations of either brain processes or cognition, but it is believed that their distributed structure allows their emergent behaviour to capture some important aspects of neurocognitive phenomena that are otherwise difficult to explain (e.g., graceful degradation and content addressability[1]). As mentioned above, in neurocomputational models the structure of a representation is defined by a large number of micro-feature dimensions (Churchland & Sejnowski, 1992). The "content" of a particular representation is specified by the array of "values" along the feature dimensions. In this perspective, the "structure" or "form" is related to the computational space defined by the feature dimensions, and this space is a platform that can be "filled" by specific contents in a particular instance. In a detailed model of cognitive schema,

[1] "Graceful degradation" refers to a pattern of loss of information in a computational system as a result of inactivation of part of its hardware components. In a conventional computer, the inactivation of a specific hardware location generally results in total loss of circumscribed information. In contrast, in distributed systems, inactivation of part of the hardware leads to global impairment affecting all information, and the initial effect of limited damage may be relatively small. "Content addressability" refers to the process of memory retrieval whereby presentation of part of the memory contents can be used to retrieve the entire memory representation. This process can be realized in some distributed networks (e.g., autoassociative networks) and is in contrast to conventional computer process of using a separate "address" to search and retrieve a piece of information.

Rumelhart, Smolensky, McClelland, and Hinton (1986) constructed a neural network where units represent features in a cognitive schema and the connection between units represents constraints between features. The network is allowed to evolve in such a way that each unit calculates its activation level based on the initial input (external evidence), according to a specified algorithm. Since the units are connected with most other units, the outputs then influence the activity level of the connected units. This process is repeated until the state of the network (i.e., the activity level of all the units) settles into a steady state. This state represents an optimal solution to a given input, in which the maximum number of internal constraints is satisfied. Rumelhart et al. (1986) illustrated this with an example using the schema for "rooms", where features of this schema would consist of size and shape of the space, the number and nature of furniture, the colour and patterns of the walls, the lighting, the inhabitants and so on. A particular permutation of final values for each of the features would then give rise to the interpretation of the kind of room encountered (e.g., whether this is a classroom, a restaurant, or a washroom). The schema for "rooms" can be filled in with specific contents (bathroom, living room, bedroom, etc.) according to external input. Importantly "contents" are not fixed entities, but are patterns specified by the strength of association between different feature dimensions, which could be adjusted based on previous encounter (e.g., according to a defined learning algorithm such as the Hebbian rule). Such systems are capable of adaptive learning, associative memory and handling ambiguous information. The contents are not actually stored; what is stored is the connection strengths, which encode the co-occurrence of features in the environment. The actualization of a particular "content" at any particular moment is therefore fluid, plastic, and adaptable.

This conceptualization has implications for the endophenotype approach: the models only need to define very few initial baseline structures, namely the feature dimensions and the computational algorithms. The actual "contents" are experience-dependent, modifiable representations. This distinction implicitly suggests that only the basic structure and the learning algorithm need to be specified biologically and are more likely to be genetically determined (hard-wired), whereas "content" can develop according to individual experience. The strength of PDP models is that they suggest innovative and interesting possibilities for key neurocomputationally relevant variables as putative endophenotypes (e.g., for size of computation network, strength of inhibitory processes, neuronal noise regulation, synaptic pruning, or synaptic learning algorithms). These variables are located between the levels of neurobiological and cognitive representations.

Neurocomputationally relevant endophenotypes are expected to be mediating processes that are generic rather than relating to specific contents.

"CONTENT" IN COGNITION AND PSYCHOPATHOLOGY

The distinction between "form (or structure)" and "content" permeates much of cognitive and psychopathological research and requires a more focused exploration here.

How "contents" have escaped cognitive models

In our consideration of cognitive scientific approaches to cognition, we have reviewed how conventional information processing approaches were limited by a preference for generic, neutral stimuli. We then reviewed recent neurocomputational approaches, which attempt further reduction for computational processes within cognitive systems, and we found that in such systems "contents" are usually represented by experience-modifiable processes rather than genetically determined processes. While suggesting interesting possibilities for neurocomputational endophenotypes (see earlier), these models attribute content-dependent processes almost entirely to learning. Although neurocomputational models themselves place much emphasis on the role and nature of learning, they have necessarily focused on modelling systems where the important "contents" of cognitions are learnt rather than inherited. The latter are relevant in endophenotype considerations. In the following sections we also review the contexts in which "contents" have also been relatively overlooked in cognitive psychology and in psychopathology.

Content and form in cognition

In conventional cognitive psychology and neuropsychology, there is a tendency towards studying how the cognitive system responds to neutral stimuli. For instance, abstract shapes, colours, and forms are used in the Wisconsin card sorting test; "meaning-free" numbers or letters are used in the continuous performance tests. The use of such stimuli originates from a convention where the target of investigation implicitly focuses on general computational processes, which are presumed to operate in a similar fashion regardless of the specific "content" of the particular stimulus. Thus in the Wisconsin card sorting test, the colour "red" is presumed to be just one of the possible colours, without particular distinctive meanings. It is not anticipated that results would differ if another colour was used in substitution for the "red" colour. Findings from such studies led to insights about how the cognitive system operates "in general" regardless of the "content" of the stimulus (content-independent processes). In other words, this approach focuses on those cognitive operations which are invariant to meaning and content. Although subsequently there are also studies which specifically investigated how particular "content" or emotional connotations may contribute towards processing, this important domain has not been emphasized and explored as intensively as the "content-independent" processes. As an example, this approach would look for whether the processing of, say the colour "red" can have an impact different from that of another colour (e.g., Elliot, Maier, Moller, Friedman, & Meinhardt, 2007). Taking this perspective, it is noteworthy that colour perception itself (as much as colour words) lead to semantic activation of concepts related to the colour (Nijboer, van Zandvoort, & de Haan, 2006). This example illustrates that much of cognitive processing is inherently content (semantics) dependent and this could affect even supposedly "neutral" information processing "tasks". Other approaches involve the "emotional Stroop task" where participants have to name the colour in which words are printed, where the words may contain contents relevant to the psychopathology. The emotional Stroop task has been extensively studied mostly in affective disorders with a general finding that words with connotations relevant to psychopathology are processed slower (Williams, Mathews, & MacLeod, 1996). Similarly, some studies of facial emotion processing suggest that the processing of specific affect might be impaired in patients with psychotic disorders (Kohler et al., 2003; Mandal, Pandey, & Prasad, 1998). While much of the data in this area are preliminary and await confirmation, these possibilities should call into question the assumption that content-independent processes should automatically be the main focus of cognitive investigations and are necessarily more biologically relevant.

Content and form in psychopathology

A similar dichotomy between "content-dependent" and "content-independent" dimensions also emerged as an important notion in descriptive psychopathology. The distinction between "content" and "form" in psychopathological phenomena can be traced back to the early descriptive psychopathologists. Karl Jaspers, in his classic textbook on psychopathology (Jaspers, 1918/1963, p. 58), remarked that the distinction between form and content is "constantly being made in psychopathology", that they should be kept separate, and that descriptive psychopathology is primarily interested in the "form" rather than the "content" of phenomena. The assumption underlying this perspective is that the "content" of a psychopathological phenomenon is more variable as it depends on individual experiences, whereas the "form" of the phenomenon may have less individual variations (e.g., a hallucination is an abnormal perception regardless of the actual content, such as what the voices actually said), and is presumably more related to the same abnormal brain processes that generate the experience in different individuals. This emphasis on "form", or representational structure, has significantly impacted descriptive psychopathology and the methodology of symptom characterization, which became the foundation for modern definitions of diagnostic categories. Notwithstanding the above general principle it is noteworthy that consideration of content has been an important aspect of descriptive psychopathology (e.g., in the classification of delusions, and when considering mood-congruity in psychotic symptoms), and it has increasingly been recognized that some aspects of "content" are consistent across subjects and the specific "contents" are themselves interesting clues for the nature of these symptoms.

Content and cognitive modules

In considering the extent to which cognitive subsystems are involved in content-dependent processing, Pylyshyn (1984) makes the distinction between subsystems that operate at the semantic, intentional level (i.e., content-dependent) from others. According to this framework, content-dependent processing is proposed to be located at a "higher" level (Pylyshyn, 1984). More specifically, the theoretical notion of "cognitive penetrability" has been introduced to investigate whether a given phenomenon is independent of "high-level" beliefs and goals. If changes in contents alter the empirical phenomenon, the phenomenon is said to be "cognitively penetrable" (Figure 2).

This approach articulates some assumptions often made only implicitly in cognitive science, namely the tendency to divide cognition into "lower level" mechanistic processes on one hand (strongly modular subsystems, presumably closer to biology, mechanical, and content-independent), and "higher level" processing on the other hand (weakly modular subsystems, presumably influenced by individual experience and culture, hence content-sensitive). In the context of this dichotomy, it is expected that research efforts may underemphasize the role of biological determination in "higher-level", content-dependent behaviours and illnesses.

RELEVANCE OF CONTENT-SPECIFIC COGNITIVE SUBSYSTEMS

The need to consider content-specific cognitive subsystems

Recent developments in cognitive science have however challenged the assumption that biologically important subsystems are necessarily content-independent (Duchaine, Cosmides, & Tooby, 2001). These authors challenged the general information processing model in which psychology tended

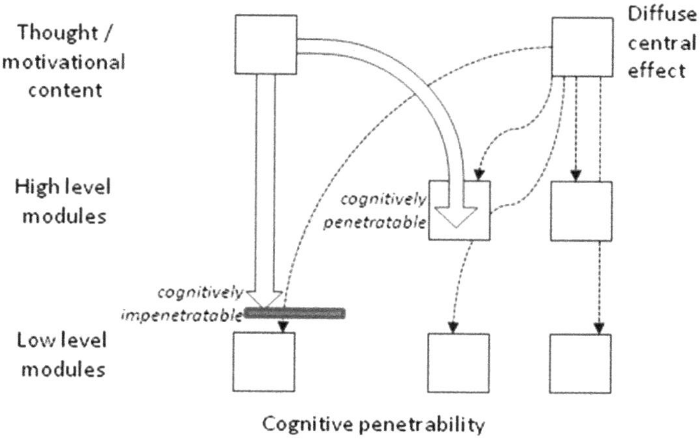

Figure 2. Cognitive penetrability and postulated hierarchy in the cognitive system. According to Pylyshyn (1984), thought/motivational content would affect only high-level modules, whereas low-level modules are unaffected, or cognitively impenetratable, by these contents. This model is shown on the left. An example of cognitive impenetrable phenomena would be the Müller-Lyer visual illusion: higher-level knowledge about the illusion does not cause it to disappear. To the right of the figure an alternative model is depicted where thought/motivational contents exert diffuse central effects directly on both presumably "high" and "low" level modules.

to withdraw from explaining meaning-laden content-dependent phenomena. It was noted that cognitive experiments have been designed mostly to explore general, content- independent mechanisms, and "emotionally-laden" stimuli have often been regarded as "noise" in experiments (Tooby & Cosmides, 1992). They argued content-independent cognitive processes as only a weak explanation since most of human cognition and psychology will need to address content/domain-dependent processes. A growing body of evidence suggests that "contents" are not simply learned combinations of features expressed in a neutral biological hardware platform. In many respects, processing of everyday stimuli is far from neutral and is often dependent on the content. As a specific example, individuals perform logical operations in a significantly different manner depending on whether abstract symbols or real-life social scenarios are used in logically and structurally equivalent tasks that differ only in "content" (Cosmides, 1989). An example of this phenomenon is the Wason selection task (Figure 3). When presented with the letter–number version of the task (Figure 3a), only a small proportion of healthy participants are able to respond correctly in this task. However, when an alternative, "social interaction" task is presented (Figure 3b), most participants would be able to correctly respond.

Based on this and other findings, it has been proposed that the human cognitive system has evolved to solve specific problems (e.g., interactions of

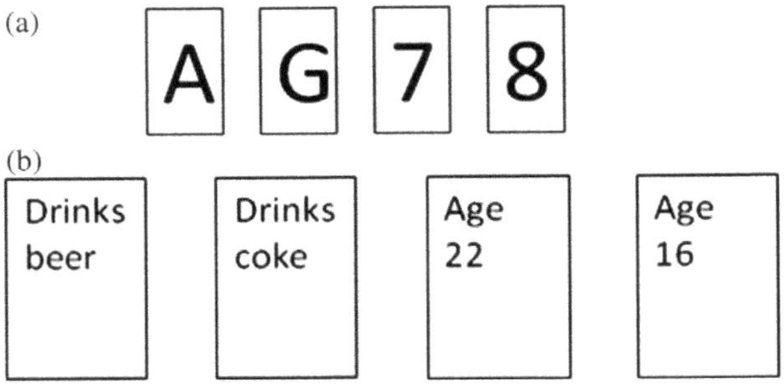

Figure 3. The Wason selection task. (a) Is a letter–number version of the task. Four cards are presented, each with a number on one side and a letter on the other. Two cards are presented letter-side up and two cards are presented number-side up. The subject is then presented with a rule: "If there is an A on one side, then there is a 7 on the other side", and then asked which cards they must turn in order to discover whether the rule is true. The correct answer is "A" and "8". (b) Is the social interaction version of the same task. The rule is "if a person is drinking a beer, then they must be over 18". The subject is asked which cards need to be turn to know if the rule has been violated.

a social nature) rather than for general abstract computation (Tooby & Cosmides, 1992). Proponents of this view suggest that the human brain handles information not in an abstract and neutral manner, but instead, in a manner relevant to potential evolutionary scenarios that early *Homo sapiens* encountered in their natural and social environment (Barkow, Cosmides, & Tooby, 1992). These observations implicitly suggest some contents will be more relevant than others with regards survival value in the environment of early hominids. Evolutionary pressure may have acted on the selection for some content-dependent cognitive functions resulting in genetically determined "hard-wiring" of the brain. For instance, returning to the example of the concept of rooms mentioned above, evolutionary pressure would have favoured a cognitive system that can discern specific types of spatial environment (e.g., an enclosed space from which there is no easy escape route, which maps closely to the stimulus that could trigger certain forms of agoraphobia).

Genetic determination of content-dependent behaviour

The biological organization of behaviours in terms of gene actions is unlikely to follow a simple hierarchical pattern. From animal studies, it is evident that certain highly complex behaviours can simply and directly map onto a small number of specific regulating genes (e.g., Drosophila sexual behaviour) (Manoli, Meissner, & Baker, 2006). Such behaviours may become highly conserved in evolution because of intense selection pressure (Hirth & Reichert, 1999). Similarly in humans, there may be behavioural sets that are under considerable selection pressure and thus are highly conserved even at a complex behavioural level. Highly conserved processes may interact with specific stimuli in the environment and constitute content-dependent rather than general processing. Successful identification of such systems requires considerations from an evolutionary biological perspective. Thus, when considering endophenotypes for a specific behavioural set, it is important for investigators to take into account whether the "black box" is likely to include content-dependent elements. Some ideas about this can be generated from consideration of the highest level phenomena (i.e., symptoms and diagnostic categories).

Content-dependence in symptom and behaviour

Illnesses can be the result of dysfunction in either content-dependent or content-independent processing. For instance, most phobias are content-dependent phenomena elicited by a highly specific set of stimuli (e.g., an enclosed space with no escape, or small insects), whereas in generalized anxiety disorder, anxiety symptoms are experienced by patients "out of the blue" and do not appear to be restricted to specific stimuli (Stevens & Price, 2000). It is important to recognize that in phobia, a specific and narrow range of stimuli (usually small animals such as insects, but seldom large dangerous animals) are found in the majority of cases, which suggests that a general "learned threats" explanation is not adequate. For many patients with animal phobias, the exposure to the phobic stimulus is minimal or sometimes even absent. On the other hand, conditions such as dementia and learning disabilities appear content-independent, as they disrupt the processing of a wide range of information regardless of content. In contrast, many other disorders, such as obsessive compulsive disorder, affective disorders, as well as psychotic disorders, are content-dependent, in which symptoms are expressed in well-circumscribed and specific sets of themes. Adequate theories of these disorders must include an account of why particular theme contents are prominent.

With the general understanding that behaviour consists of a wide spectrum of adaptive components, it has often been assumed that responses to specific environmental stimuli are adaptively acquired (learned) and experience-dependent instead of biologically determined. This presumption has led to the tendency to consider content-dependent processes primarily in terms of cultural and social contexts, resulting in an overemphasis of content-independent processes in neurobiological illness theories and relative neglect of the potential biological underpinning and significance of content-dependent processes.

As content-dependent processing may constitute an important aspect of the cognitive system and may be of direct evolutionary biological relevance, it follows that potentially fruitful lines of enquiry may be suggested if human brain function and psychopathology are explored in the context where they have evolved to confront specific challenges for survival in unique physical and social environments, rather than as an all-purpose general information processing system (Barkow et al., 1992; Black, 1998). Application of an evolutionary framework in understanding the adaptive functions of symptoms and cognitive processes have been proposed in some areas of psychopathology. For instance, in an interesting discussion on the possible endophenotypes of dysthymia from an evolutionary perspective, it has been proposed that an "anergic" form of dysthymia, with the symptoms of loss of motivation, loss of interest, and loss of engagement with daily tasks, may have evolutionary selective advantage in the sense that it allows the individual to withdraw from a challenging situation where persistent engagement may not be the most advantageous strategy (Niculescu & Akiskal, 2001).

Given that some content-dependent disorders may have a prominent genetic basis (e.g., psychosis), it would be an important strategy to search for genes that map to content-dependent behavioural phenotypic expression, even though such systems may be less accessible and currently less well-defined. In contrast, while content-independent processes (e.g., general information processing capacity) may contribute to the explanation, they are on their own unlikely to provide adequate accounts for these disorders. To ignore content-dependence in cognitive processes and to focus exclusively on the more accessible content-independent endophenotypes as an account for content-dependent disorders is a mismatch that could compromise the eventual explanatory power.

Need to reconsider psychopathological endophenotypes in addition to cognitive endophenotypes

Before we embark on the challenge of exploring content-dependent cognitive processes underlying psychosis, it should be asked whether content-dependent approaches have been fully exploited at the level of symptom specifications. Current endophenotype research seldom capitalizes on symptoms as a possible level of analysis. Some accounts are even explicitly excluding symptoms in their consideration of endophenotypes, perhaps out of the belief that they are too closely related to the clinical categories and too complex or "experiential" (a notable exception is found in Gottesman & Gould, 2003, where subjective data were included as possible endophenotypes). Often research approaches bypass the level of symptom expression and attempt to relate clinical categories with more distal neurocognitive and neurophysiological processes. Symptoms however remain one of the most content-dependent expressions of the disorder. This bypassing is potentially costly as it immediately reduces the clinical phenomenon of psychosis to something far less content-dependent (i.e., that of a general information processing deficit).

Arguably the study of symptoms and their structures should be a prime target of endophenotype research. At present, the complexity and descriptive methodology of symptoms may have been under-investigated. Although there are known possibilities of variability and measurement error in symptoms, given sufficient attention to methodology, these are not necessarily insurmountable. As discussed above, it is often assumed that the form of a symptom is more invariant than the content of a symptom. However, form variables are insufficient and may constraint an adequate account of psychopathology. For example, they fall short in explaining why delusions are often on a limited range of themes, and why hallucinations are usually verbal and derogatory. Theoretical explanations that account for the content of psychotic symptoms (content-relevant theories) are critical in understanding psychosis.

Because of the general tendency in cognition research to neglect specific content information contained in symptoms, current approaches may grossly understudy the richness in the psychosis phenomena. In addition, even for the content-independent symptoms, the current perspectives may be suboptimal. Psychiatric research, in the search for reliable diagnostic categories, may have trivialized the importance of symptom identification, characterization, and classification. It is often assumed that the ascertainment and characterization of symptoms are unproblematic, and focus has been more on how combinations of symptoms produce diagnostic categories (Berrios & Chen, 1993). In fact, the characterization of symptoms may represent a central challenge to psychiatric research in general, and to the endophenotype approach in particular. The current under-emphasis on symptoms has resulted in the widespread use of pragmatic symptom rating scales that may lack the sensitivity to measure potentially important symptom dimensions, as well as the capacity to capture subthreshold phenomena in non-clinical populations. This has resulted in a relatively unrefined symptom characterization, foregoing the opportunity to explore to what extent symptom dimensions reflect underlying neurobiological system dysfunctions.

CONCLUSIONS: BIOLOGICALLY RELEVANT ENDOPHENOTYPES

Based on the above discussion, it seems reasonable to expect endophenotypes for psychosis (and any content-dependent disorders) to be of two important classes: generic cognitive endophenotypes and content-dependent endophenotypes. Generic endophenotypes are related to general cognitive performance and capacity and they are content-neutral. It is expected that generic endophenotypes may be shared by many different disorders (i.e., they may determine the general risk for a variety of mental conditions). There is a need for an optimal framework for studying these content-independent processes. In-depth consideration of cognitive processes (such as neurocomputational processes) may provide further insights and possibilities. In contrast, content-dependent endophenotypes are those that act to determine the content-dependent expression of disorders (such as psychotic symptoms). They account for the specific disorders expressed. Content-dependent endophenotypes, which are relatively under-explored due to relative lack of available measurements tools, are expected to be linked to evolutionary-relevant biological determinants of behaviour. Investigation of content-dependent processes and abnormalities may constitute one of the greatest challenges and opportunities in future psychosis research.

REFERENCES

Barkow, J. H., Cosmides, L., & Tooby, J. (1992). *The adapted mind: Evolutionary psychology and the generation of culture.* New York: Oxford University Press.

Berrios, G. E., & Chen, E. Y. (1993). Recognising psychiatric symptoms. Relevance to the diagnostic process. *British Journal of Psychiatry, 163*, 308–314.

Black, I. B. (1998). Genes, brain, and mind: The evolution of cognition. *Neuron, 20*(6), 1073–1080.

Butcher, L. M., Kennedy, J. K., & Plomin, R. (2006). Generalist genes and cognitive neuroscience. *Current Opinion of Neurobiology, 16*(2), 145–151.

Chen, E. Y. H. (1994). A neural-network model of cortical information-processing in schizophrenia .1. Interaction between biological and social-factors in symptom formation. *Canadian Journal of Psychiatry-Revue Canadienne de Psychiatrie, 39*(8), 362–367.

Chen, E. Y. H. (1995). A neural-network model of cortical information-processing in schizophrenia .2. Role of hippocampal-cortical interaction – A review and a model. *Canadian Journal of Psychiatry-Revue Canadienne de Psychiatrie, 40*(1), 21–26.

Churchland, P. S., & Sejnowski, T. J. (1992). *The computational brain.* Cambridge, MA: MIT Press.

Cosmides, L. (1989). The logic of social exchange: Has natural selection shaped how humans reason? Studies with the Wason selection task. *Cognition, 31*(3), 187–276.

Dickinson, D., Ragland, J. D., Calkins, M. E., Gold, J. M., & Gur, R. C. (2006). A comparison of cognitive structure in schizophrenia patients and healthy controls using confirmatory factor analysis. *Schizophrenia Research, 85*(1–3), 20–29.

Duchaine, B., Cosmides, L., & Tooby, J. (2001). Evolutionary psychology and the brain. *Current Opinion in Neurobiology, 11*(2), 225–230.

Elliot, A. J., Maier, M. A., Moller, A. C., Friedman, R., & Meinhardt, J. (2007). Color and psychological functioning: The effect of red on performance attainment. *Journal of Experimental Psychology: General, 136*(1), 154–168.

Flint, J., & Munafo, M. R. (2007). The endophenotype concept in psychiatric genetics. *Psychology and Medicine, 37*(2), 163–180.

Fodor, J. A. (1983). *The modularity of mind: An essay on faculty psychology.* Cambridge, MA: MIT Press.
Goldberg, T. E., & Weinberger, D. R. (2004). Genes and the parsing of cognitive processes. *Trends in Cognitive Science, 8*(7), 325–335.
Goldberg, T. E., Weinberger, D. R., Berman, K. F., Pliskin, N. H., & Podd, M. H. (1987). Further evidence for dementia of the prefrontal type in schizophrenia? A controlled study of teaching the Wisconsin Card Sorting Test. *Archives of General Psychiatry, 44*(11), 1008–1014.
Goldman-Rakic, P. S. (1994). Working memory dysfunction in schizophrenia. *Journal of Neuropsychiatry and Clinical Neurosciences, 6*(4), 348–357.
Gottesman, II., & Gould, T. D. (2003). The endophenotype concept in psychiatry: Etymology and strategic intentions. *American Journal of Psychiatry, 160*(4), 636–645.
Gottesman, II., & Shields, J. (1967). A polygenic theory of schizophrenia. *Proceedings of the National Academy of Science of the United States of America, 58*(1), 199–205.
Gould, T. D., & Gottesman, II. (2006). Psychiatric endophenotypes and the development of valid animal models. *Genes, Brain and Behavior, 5*(2), 113–119.
Harrison, P. J., & Owen, M. J. (2003). Genes for schizophrenia? Recent findings and their pathophysiological implications. *Lancet, 361*(9355), 417–419.
Hirth, F., & Reichert, H. (1999). Conserved genetic programs in insect and mammalian brain development. *Bioessays, 21*(8), 677–684.
Hoffman, R. E. (1987). Computer simulations of neural information processing and the schizophrenia-mania dichotomy. *Archives of General Psychiatry, 44*(2), 178–188.
Jaspers, K. (1918/1963). *General psychopathology.* In J. Hoenig & M. W. Hamilton (Eds.), (Trans. 7th ed.). Manchester: Manchester University Press.
John, B., & Lewis, K. R. (1966). Chromosome variability and geographic distribution in insects. *Science, 152*(3723), 711–721.
Keri, S., & Janka, Z. (2004). Critical evaluation of cognitive dysfunctions as endophenotypes of schizophrenia. *Acta Psychiatrica Scandinavica, 110*(2), 83–91.
Kohler, C. G., Turner, T. H., Bilker, W. B., Brensinger, C. M., Siegel, S. J., Kanes, S. J., et al. (2003). Facial emotion recognition in schizophrenia: Intensity effects and error pattern. *American Journal of Psychiatry, 160*(10), 1768–1774.
Koponen, S., Taiminen, T., Portin, R., Himanen, L., Isoniemi, H., Heinonen, H., et al. (2002). Axis I and II psychiatric disorders after traumatic brain injury: A 30-year follow-up study. *American Journal of Psychiatry, 159*(8), 1315–1321.
Kosslyn, S. M. (1994). *Image and brain: The resolution of the imagery debate.* Cambridge, MA: MIT Press.
Kovas, Y., & Plomin, R. (2006). Generalist genes: Implications for the cognitive sciences. *Trends in Cognitive Science, 10*(5), 198–203.
Kraus, M. S., & Keefe, R. S. (2007). Cognition as an outcome measure in schizophrenia. *British Journal of Psychiatry Suppl, 50*, s46–51.
MacCabe, J. H., Lambe, M. P., Cnattingius, S., Torrang, A., Bjork, C., Sham, P. C., et al. (2008). Scholastic achievement at age 16 and risk of schizophrenia and other psychoses: A national cohort study. *Psychological Medicine, 38*(8), 1133–1140.
Manoli, D. S., Meissner, G. W., & Baker, B. S. (2006). Blueprints for behavior: Genetic specification of neural circuitry for innate behaviors. *Trends in Neuroscience, 29*(8), 444–451.
Meyer-Lindenberg, A., & Weinberger, D. R. (2006). Intermediate phenotypes and genetic mechanisms of psychiatric disorders. *Nature Reviews Neuroscience, 7*(10), 818–827.
Mandal, M. K., Pandey, R., & Prasad, A. B. (1998). Facial expressions of emotions and schizophrenia: A review. *Schizophrenia Bulletin, 24*(3), 399–412.
Niculescu 3rd, A. B., & Akiskal, H. S. (2001). Proposed endophenotypes of dysthymia: Evolutionary, clinical and pharmacogenomic considerations. *Molecular Psychiatry, 6*(4), 363–366.

Nijboer, T. C., van Zandvoort, M. J., & de Haan, E. H. (2006). Seeing red primes tomato: Evidence for comparable priming from colour and colour name primes to semantically related word targets. *Cognitive Processing, 7*(4), 269–274.

Page, G. P., George, V., Go, R. C., Page, P. Z., & Allison, D. B. (2003). "Are we there yet?": Deciding when one has demonstrated specific genetic causation in complex diseases and quantitative traits. *American Journal of Human Genetics, 73*(4), 711–719.

Pylyshyn, Z. W. (1984). *Computation and cognition: Toward a foundation for cognitive science.* Cambridge, MA: MIT Press.

Rosen, J., & Zubenko, G. S. (1991). Emergence of psychosis and depression in the longitudinal evaluation of Alzheimer's disease. *Biology and Psychiatry, 29*(3), 224–232.

Rumelhart, D. E., & McClelland, J. L. (1986). *Parallel distributed processing: Explorations in the microstructure of cognition.* Cambridge, MA: MIT Press.

Rumelhart, D. E., Smolensky, P., McClelland, J. L., & Hinton, G. E. (1986). Schemata and sequential thought process in PDP models. In D. E. Rumelhart & J. L. McClelland (Eds.), *Parallel distributed processing: Explorations in the microstructure of cognition* (Vol. 2, pp. 7–57). Cambridge, MA: MIT Press.

Saykin, A. J., Shtasel, D. L., Gur, R. E., Kester, D. B., Mozley, L. H., Stafiniak, P., et al. (1994). Neuropsychological deficits in neuroleptic naive patients with first-episode schizophrenia. *Archives of General Psychiatry, 51*(2), 124–131.

Simon, H.A. (1981). *The sciences of the artificial.* Cambridge, MA: MIT Press.

Smolensky, P. (1995). On the proper treatment of connectionism. In C. MacDonald & G. MacDonald (Eds.), *Connectionism: Debates on psychological explanation* (pp. 28–89). Oxford: Blackwell Publishers.

Stevens, J., & Price, J. (2000). *Evolutionary psychiatry: A new beginning* (2nd ed.). London: Routledge.

Tooby, J., & Cosmides, L. (1992). The psychological foundations of culture. In J. H. Barkow, L. Cosmides & J. Tooby (Eds.), *The adapted mind: Evolutionary psychology and the generation of culture* (pp. xii, 666). New York: Oxford University Press.

Toulopoulou, T., Picchioni, M., Rijsdijk, F., Hua-Hall, M., Ettinger, U., Sham, P., et al. (2007). Substantial genetic overlap between neurocognition and schizophrenia: Genetic modeling in twin samples. *Archives of General Psychiatry, 64*(12), 1348–1355.

Turner, T. H. (1989). Schizophrenia and mental handicap: An historical review, with implications for further research. *Psychology and Medicine, 19*(2), 301–314.

van Os, J., Howard, R., Takei, N., & Murray, R. (1995). Increasing age is a risk factor for psychosis in the elderly. *Social Psychiatry and Psychiatric Epidemiology, 30*(4), 161–164.

Williams, J. M., Mathews, A., & MacLeod, C. (1996). The emotional Stroop task and psychopathology. *Psychology Bulletin, 120*(1), 3–24.

For Product Safety Concerns and Information please contact our EU
representative GPSR@taylorandfrancis.com
Taylor & Francis Verlag GmbH, Kaufingerstraße 24, 80331 München, Germany

www.ingramcontent.com/pod-product-compliance
Ingram Content Group UK Ltd.
Pitfield, Milton Keynes, MK11 3LW, UK
UKHW020156260926
475702UK00047B/298

* 9 7 8 1 8 4 8 7 2 7 1 8 2 *